Contents

Notations   1

§ 1. Introduction   3

§ 2. Lemmas   7

§ 3. Theorems with positive or decreasing functions 13

§ 4. Theorems with positive or decreasing coefficients 19

§ 5. The exceptional integral values of the index 26

§ 6. L^p problems, $1 < p < \infty$ 35

§ 7. Asymptotic formulas and Lipschitz conditions 43

§ 8. More general classes of functions; conditional convergence 52

§ 9. Trigonometric integrals 57

Bibliography 60

Index 64

Prof. Dr. Ralph P. Boas, Jr.
Northwestern University
Evanston, Illinois, U.S.A.

© by Springer-Verlag Berlin · Heidelberg 1967

Library of Congress Catalog Card Number 67–18963
Printed in Germany

Titel-Nr. 4582

Ralph P. Boas, Jr.

Integrability Theorems
for Trigonometric Transforms

Springer-Verlag Berlin Heidelberg New York 1967

Ergebnisse der Mathematik
und ihrer Grenzgebiete

Band 38

Herausgegeben von

P. R. Halmos · P. J. Hilton · R. Remmert · B. Szőkefalvi-Nagy

Unter Mitwirkung von

L. V. Ahlfors · R. Baer · F. L. Bauer · R. Courant · A. Dold
J. L. Doob · E. B. Dynkin · S. Eilenberg · M. Kneser · M. M. Postnikov
H. Rademacher · B. Segre · E. Sperner

Geschäftsführender Herausgeber: P. J. Hilton

Notations

We are concerned with formal sine and cosine series

$$g(x) \sim \sum_{n=1}^{\infty} b_n \sin nx,$$

$$f(x) \sim \tfrac{1}{2} a_0 + \sum_{n=1}^{\infty} a_n \cos nx,$$

and we preserve this notation throughout: a_n are always (possibly formal) cosine coefficients associated with f; b_n are sine coefficients associated with g; similarly for F and A_n, G and B_n. The functions f and g are understood (except in §9) to be real and to have domain $(0, \pi)$. By φ we mean either an f or a g, in cases where it makes no difference which one we are talking about; the coefficients associated with φ are denoted by λ_n (with the convention that $b_0 = 0$; but "$\{b_n\}$ decreases," for example, means that $\{b_n\}$ decreases from b_1 on).

A Fourier sine or cosine series is understood to mean the Fourier series of a (Lebesgue) integrable function. When g is not necessarily integrable but $x\, g(x)$ is, we call

$$b_n = 2\,\pi^{-1} \int_0^{\pi} g(x) \sin nx \, dx$$

the generalized sine coefficients of g. This is standard terminology. We shall also need generalized cosine coefficients, defined, when $x^2 f(x)$ is integrable, by

$$-a_n = 2\,\pi^{-1} \int_0^{\pi} f(x)\,(1 - \cos nx)\, dx.$$

"Positive" and "increasing" mean "nonnegative" and "nondecreasing;" similarly for "negative" and "decreasing." The symbols $\uparrow$, $\downarrow$ stand for "increases," "decreases," "increases to," "increasing," and the like.

"Ultimately," as in "$a_n \geq 0$ ultimately," means "for all sufficiently large n."

The letter C, possibly with subscripts, denotes a number independent of everything except the subscripts; a C may stand for a different number from one appearance to another.

Sums with unspecified limits are over $[1, \infty)$; sums with nonintegral limits are over the integers between the limits.

Integrals with unspecified limits are over $(0, \pi)$.

$$\int_{a+}^{x} \quad \text{means} \quad \lim_{\varepsilon \to 0+} \int_{a+\varepsilon}^{x}.$$

$$\Delta a_n = a_n - a_{n+1}.$$

If $p > 1$, $p' = p/(p-1)$.

A theorem or proof in which the convergence of a series implies the convergence of an integral is often identified by the symbol $\sum \to \int$; one in which the implication goes the other way, by $\int \to \sum$.

Preface

This monograph is a report on the present state of a fairly coherent collection of problems about which a sizeable literature has grown up in recent years. In this literature, some of the problems have, as it happens, been analyzed in great detail, whereas other very similar ones have been treated much more superficially. I have not attempted to improve on the literature by making equally detailed presentations of every topic. I have also not aimed at encyclopedic completeness. I have, however, pointed out some possible generalizations by stating a number of questions; some of these could doubtless be disposed of in a few minutes; some are probably quite difficult.

This monograph was written at the suggestion of B. Sz.-Nagy. I take this opportunity of pointing out that his paper [1] inspired the greater part of the material that is presented here; in particular, it contains the happy idea of focusing attention on the multipliers $n^{\gamma-1}$, $x^{-\gamma}$. R. Askey, P. Heywood, M. and S. Izumi, and S. Wainger have kindly communicated some of their recent results to me before publication. I am indebted for help on various points to L. S. Bosanquet, S. M. Edmonds, G. Goes, S. Izumi, A. Zygmund, and especially to R. Askey. My work was supported by the National Science Foundation under grants GP-314, GP-2491, GP-3940 and GP-5558.

Evanston, Illinois, February, 1967

R. P. Boas, Jr.

§ 1. Introduction

Suppose that a periodic function f is associated with a trigonometric cosine series $\frac{1}{2} a_0 + \sum a_n \cos nx$; under various circumstances we may want to suppose that f is integrable (in some sense) and that a_n are its Fourier coefficients (in some sense), or that the series converges (in some sense) to the function. We ask two questions: (a) if ψ is a given positive function, and f belongs to a specified class of functions, what hypotheses on $\{a_n\}$ are equivalent to $f \psi \in L$? (b) If $\{\mu_n\}$ is a given sequence of positive numbers, and $\{a_n\}$ belongs to a given class of sequences, what hypotheses on f are equivalent to $\sum \mu_n |a_n| < \infty$? Parseval's theorem suggests that the answers should have something to do with $\sum \mu_n a_n$ and $\int f \psi$, respectively, where μ_n are the cosine coefficients of ψ. However, Parseval's equation, even supposing it to be true for the functions and coefficients concerned, will not show absolute integrability or absolute convergence unless the functions or coefficients are positive.

When $0 < \gamma < 1$, a function that behaves near 0 like $x^{-\gamma}$ has a cosine series whose coefficients behave at infinity like $n^{\gamma-1}$, and conversely. This convenient fact dominates most of the theory, so that we shall be concerned primarily with conditions for the convergence of $\sum |a_n| n^{\gamma-1}$ or of $\int |f(x)| x^{-\gamma} d x$ — not, however, always for $0 < \gamma < 1$. The generally elegant appearance of the theorems depends on assuming that either $f(x)$ or $\{a_n\}$ is positive, or monotonic, or something of the sort. To illustrate the general pattern, we state a number of theorems in their most symmetric form[1]).

Theorem 1. (§ 3) *If $f(x) \downarrow$ on $(0, \pi)$, $f \in L$, and f is bounded below, and $0 < \gamma < 1$, then $x^{-\gamma} f(x) \in L$ if and only if $\sum |a_n| n^{\gamma-1}$ converges.*

The general symmetry between a function and its Fourier coefficients suggests a dual theorem in which the roles of the function and the coefficients are interchanged:

If $a_n \downarrow 0$ (in which case $\sum a_n \cos nx$ converges on $0 < x < \pi$ to a continuous $f(x)$), and $0 < \gamma < 1$, then $\sum a_n n^{-\gamma}$ converges if and only if $f(x) x^{\gamma-1} \in L$.

If we now replace γ by $1-\gamma$ we get a theorem that looks just like Theorem 1 except that the main hypothesis is imposed on the coefficients instead of the function.

Theorem 2. (§ 4) *If $a_n \downarrow 0$ and $0 < \gamma < 1$, then $x^{-\gamma} f(x) \in L$ if and only if $\sum a_n n^{\gamma-1}$ converges.*

---

[1]) The statement of each theorem includes a note of the section where it is discussed in more detail, and where references to its source will be found.

The question now arises whether Theorems 1 and 2 can be extended to other values of γ. The values 0 and 1 are exceptional, and will be discussed in § 5. At first sight we might expect no extension, either to $\gamma < 0$ or to $\gamma > 1$. In Theorem 1, for example, $x^{-\gamma} f(x) \in L$ says no more than $f \in L$ when $\gamma < 0$; in Theorem 2, if $\gamma > 1$, $\sum a_n$ converges, f is uniformly continuous with $f(0) > 0$, and $x^{-\gamma} f(x)$ cannot be integrable.

However, these failures happen for rather superficial reasons. In Theorem 1 we could try the effect of requiring only that $x^2 f(x) \in L$ and letting a_n be generalized Fourier coefficients; in Theorem 2 we could reasonably ask whether $x^{-\gamma}[f(x) - f(0)]$ is integrable, since it is integrable except at 0, and might not be integrable at 0 when $\gamma \geq 1$. It turns out that both kinds of generalizations are possible, even with a weakening of the hypothesis that the function or the sequence of coefficients is monotonic.

Theorem 3. (§ 3) *If $f(x) \geq 0$ near 0, $x^2 f(x) \in L$, f is bounded below, a_n are the generalized cosine coefficients of f, and $-2 < \gamma < 0$, then $x^{-\gamma} f(x) \in L$ if and only if $\sum |a_n| n^{\gamma - 1}$ converges.*

Theorem 4. (§ 4) *If a_n are the Fourier cosine coefficients of f, f is continuous at 0, $a_n \geq 0$, and $1 < \gamma < 3$, then $x^{-\gamma}[f(x) - f(0)] \in L$ if and only if $\sum a_n n^{\gamma - 1}$ converges.*

If $a_n \geq 0$, $\sum a_n \cos nx$ does not necessarily converge; the assumption that it is a Fourier series is natural in the context of Theorem 4 since when $\sum a_n n^{\gamma - 1}$ converges, $\sum a_n \cos nx$ even converges uniformly.

An alternative way of stating Theorem 4 is to suppose that $a_n \geq 0$ ultimately, $\frac{1}{2} a_0 + \sum a_n = 0$ (so that $f(0) = 0$), and consider the condition $x^{-\gamma} f(x) \in L$.

Other nonintegral values of γ could be used if the sequence of Fourier coefficients or the function is still further modified.

Now there is a general principle that a Fourier series with positive coefficients tends to behave about as well at all points as it does at 0[1]. When $a_n \downarrow 0$ and $0 < \gamma < 1$ (Theorem 2), $|x - a|^{-\gamma}[f(x) - f(a)]$ is trivially integrable for $0 < a < \pi$. However, when $a_n \geq 0$ and $f \in L$, this is no longer necessarily true. Hence it makes sense to ask, in Theorem 4, whether $|x - a|^{-\gamma}[f(x) - f(a)]$ is integrable when $a \neq 0$. On the other hand, if $\gamma \geq 2$ and $\sum a_n n^{\gamma - 1}$ converges, $f'(x)$ exists and is continuous, and so $|x - a|^{-\gamma}[f(x) - f(a)]$ cannot be integrable for all a unless f is a constant. These considerations lead to the following theorem.

Theorem 5. (§ 4) *If a_n are the Fourier cosine coefficients of the continuous function f, if $a_n \geq 0$, and if $1 < \gamma < 2$, then $|x - a|^{-\gamma}[f(x) - f(a)] \in L$ for $0 < a < \pi$, if and only if $\sum a_n n^{\gamma - 1}$ converges.*

Hence when $a_n \geq 0$ it is indeed true that $|x - a|^{-\gamma}[f(x) - f(a)] \in L$ for all a if this holds for $a = 0$, provided $1 < \gamma < 2$.

---

[1] A striking example of the failure of this principle has recently been found by S. WAINGER: there are Fourier series with positive coefficients, belonging to $L^p (1 < p < 2)$ near 0 but not on $(-\pi, \pi)$.

Next we ask whether the theorems can be extended from L to L^p, $p > 1$. Theorems 1, 2, 4 and 5 have the following extensions.

Theorem 6. (§ 6) *If $f(x)$ is positive, integrable, and decreasing, $p > 1$, and $-1/p' < \gamma < 1/p$, then $\sum n^{-\gamma p} a_n^p$ converges if and only if $x^{p\gamma+p-2} f(x)^p \in L$. If $a_n \downarrow 0$, $p > 1$, and $-1/p' < \gamma < 1/p$, then $x^{-\gamma} f(x) \in L^p$ if and only if $\sum a_n^p n^{p+p\gamma-2}$ converges.*

Theorem 7. (§ 6) *If a_n are the Fourier cosine coefficients of the continuous function f, if $a_n \geq 0$, and if $1/p < \gamma < (1/p)+2$, then $x^{-\gamma}[f(x)-f(0)] \in L^p$ if and only if*

$$\sum n^{p\gamma-2} \left(\sum_{k=n}^{\infty} a_k \right) p < \infty.$$

If $1/p < \gamma < (1/p)+1$, we can replace $x^{-\gamma}[f(x)-f(0)] \in L^p$ by the condition $|x-a|^{-\gamma}[f(x)-f(a)] \in L^p$ for $0 < a < \pi$.

Note that the series conditions in Theorem 7 and the second part of Theorem 6 both go over to $\sum a_n n^{\gamma-1} < \infty$ when $p = 1$.

It is natural to try the effect of letting $p \to \infty$ in Theorems 6 and 7. Formally, the conditions of the second part of Theorem 6 go over to $f(x) = O(x^\gamma)$, $a_n = O(n^{-\gamma-1})$, $-1 < \gamma < 0$; those of Theorem 7, to $f(x)-f(a) = O(|x-a|^\gamma)$, $\sum_{k=n}^{\infty} a_k = O(n^{-\gamma})$, $0 < \gamma < 1$. The theorems obtained in this way are in fact correct; there are also more precise conclusions under more precise hypotheses. It is convenient to write $\alpha = \gamma+1$.

Theorem 8. (§ 7) *If $a_n \downarrow 0$ and $0 < \alpha < 1$, then $f(x) = O(x^{\alpha-1})$, $x \to 0+$, if and only if $a_n = O(n^{-\alpha})$; $f(x) \sim A x^{\alpha-1}$, $x \to 0+$, if and only if*

$$a_n \sim (2/\pi) A \cos \tfrac{1}{2}\pi\alpha \, n^{-\alpha}.$$

Theorem 9. (§ 7) *If a_n are the Fourier cosine coefficients of f, and $a_n \geq 0$, then $f \in \text{Lip } \gamma \, (0 < \gamma < 1)$ if and only if*

$$\sum_{k=n}^{\infty} a_k = O(n^{-\gamma}).$$

If also $a_n \downarrow 0$ the last condition is equivalent zo $a_n = O(n^{-\gamma-1})$.

Theorem 10. (§ 7) *If a_n are the Fourier coefficients of f, $a_n \geq 0$, and $1 < \alpha < 2$, then $f(x) - f(0) \sim A x^{\alpha-1}$ if and only if*

$$\sum_{j=0}^{\infty} a_{2j+n+1} \sim (A/\pi) \Gamma(\alpha-1) \cos \tfrac{1}{2}\pi\alpha \, n^{1-\alpha}.$$

If $a_n \downarrow 0$ the last condition is equivalent to

$$a_n \sim (2 A/\pi) \Gamma(\alpha) \cos \tfrac{1}{2}\pi\alpha \, n^{-\alpha}.$$

Theorem 9 is the extension of the first part of Theorem 8 to $1 < \alpha < 2$, and Theorem 10 is the corresponding extension of the asymptotic formula in Theorem 9.

There is a similar set of theorems for sine series. The theorems corresponding to Theorems 1 and 2 are in fact cases of Parseval's theorem, because the sine series of a positive decreasing function has positive coefficients and a sine series with coefficients decreasing to 0 represents a function that is positive in a right-hand neighborhood of 0. Another difference is that $\gamma = 1$ is not an exceptional index for sine series because the convergence of $\sum b_n$ to a positive sum does not interfere with the integrability of $x^{-\gamma} g(x)$ for $1 \leq \gamma < 2$. Similarly, $\gamma = 1/p$ is not exceptional in the sine analogue of Theorem 6, and $\alpha = 1$ is not exceptional in the sine analogue of Theorem 8.

Sine and cosine theorems with indices γ differing by 1 are more or less equivalent. To see how this comes about, suppose for example that

$$a_n = 2\pi^{-1} \int f(x) \cos nx \, dx$$

and that f is absolutely continuous. Then

$$b_n = -n a_n = 2\pi^{-1} \int f'(x) \sin nx \, dx = 2\pi^{-1} \int g(x) \sin nx \, dx.$$

Consequently if the convergence of $\sum n^{\gamma-1}|b_n|$ is necessary and sufficient for the integrability of $x^{-\gamma} g(x)$ when $g(x) \geq 0$, then the convergence of $\sum n^{(\gamma+1)-1}|a_n|$ is necessary and sufficient for the integrability of $x^{-\gamma} f'(x)$, and hence for the integrability of $x^{-(\gamma+1)} f(x)$, when $f(x)$ decreases and is bounded below (and is absolutely continuous; this condition is easily removed); and conversely. In other words, a sine theorem with a positive function is equivalent to a cosine theorem with a decreasing function and index increased by 1.

In a similar, but technically more complicated, way, partial summation shows that a sine theorem with positive coefficients is equivalent to a cosine theorem with decreasing coefficients and index *decreased* by 1. (The exceptional behavior of integral powers under differentiation and integration accounts for the special role of integral values of γ.) Hence in principle it is best to prove our theorems with the weakest possible "positivity" condition for γ as close to zero as we can manage, and extend them to other values of γ by partial integration or partial summation.

In §§ 2−7 we shall give proofs of Theorems 1−10 and of their analogues for sine series, in most cases in a somewhat generalized form; and we shall discuss a number of cognate results. In §§ 8 and 9 we discuss briefly some generalizations in other directions. The powers $x^{-\gamma}$, $n^{\gamma-1}$ can be replaced by more general functions, absolute convergence can be replaced by conditional convergence, and monotonic sequences can be replaced by quasi-monotonic sequences (§ 8). Trigonometric series can be replaced by trigonometric integrals (§ 9). There are also a number of extensions, not discussed here, to series of other orthogonal functions: ultraspherical polynomials (ASKEY and WAINGER [2], [3], GANSER [1]), Jacobi polynomials (ASKEY [2]), Haar functions (GOLUBOV [1], LEINDLER [2]), and Walsh functions (ŠNEĬDER [1], KOKILAŠVILI [1]). Generalizations to higher dimensions have hardly been touched; one theorem has been given by TELJAKOVSKIĬ [1].

§ 2. Lemmas

We collect here some frequently used lemmas about partial summation and special trigonometric series. First we have a variant of partial summation involving the differences of terms whose indices differ by 2; this frequently (not always) simplifies formulas for trigonometric series.

Lemma 2.1.

$$\sum_{k=1}^{n} x_k(y_{k+1} - y_{k-1}) = -x_1 y_0 - x_2 y_1 + \sum_{k=2}^{n-1} y_k(x_{k-1} - x_{k+1}) + y_n x_{n-1} + y_{n+1} x_n.$$

We shall want the cases when either x_k or y_k is $\cos kx$ or $\sin kx$.

Lemma 2.2.

$$(2\sin x) \sum_{k=1}^{n} a_k \cos kx = -a_2 \sin x + \sum_{k=2}^{n-1} (a_{k-1} - a_{k+1}) \sin kx +$$

$$+ a_{n-1} \sin nx + a_n \sin(n+1)x.$$

Lemma 2.3.

$$(-2\sin x) \sum_{k=1}^{n} b_k \sin kx = -b_2 \cos x - b_1 + \sum_{k=2}^{n-1} (b_{k-1} - b_{k+1}) \cos kx$$

$$+ b_{n-1} \cos nx + b_n \cos(n+1)x.$$

Lemma 2.4. *Let* $\sum a_k$ *converge and let*

$$A_k = \sum_{j=0}^{\infty} a_{2j+k+1}.$$

Then

$$\sum_{k=1}^{n} a_k \cos kx = A_0 \cos x + A_1 \cos 2x - 2 \sum_{k=2}^{n-1} A_k \sin kx \sin x$$

$$+ A_n \cos(n-1)x + A_{n+1} \cos nx.$$

Lemma 2.5. *Let* $\sum b_k$ *converge and let*

$$B_k = \sum_{j=0}^{\infty} b_{2j+k+1}.$$

Then

$$\sum_{k=1}^{n} b_k \sin kx = B_0 \sin x + 2 \sum_{k=1}^{n-1} B_k \cos kx \sin x$$

$$+ B_n \sin(n-1)x + B_{n+1} \sin nx.$$

Lemmas 2.2 and 2.3 transform cosine or sine series with decreasing coefficients into sine or cosine series with positive coefficients; Lemmas 2.4 and 2.5 do just the opposite.

The following lemma about series of numbers is a consequence of Lemma 2.1.

Lemma 2.6. *If* $x_k \downarrow 0$, $y_k \geq 0$, *and* $y_k \uparrow$, *then the series* $\sum x_k(y_{k+1} - y_{k-1})$ *and* $\sum y_k(x_{k-1} - x_{k+1})$ *converge or diverge together.*

If the first series converges, the sum on the left of Lemma 2.1 increases with n and is bounded; the sum on the right is positive; so $y_n x_{n-1} + y_{n+1} x_n$ is bounded above and is positive. Hence the second series has bounded partial sums and positive terms, and therefore it converges. Conversely, if the second series converges,

$$y_N(x_{N-1} + x_N) \leq \sum_N^\infty y_k(x_{k-1} - x_{k+1}) \to 0,$$

so $y_n x_{n-1} + y_{n+1} x_n \to 0$; then the right-hand side of Lemma 2.1 approaches a limit, so the left-hand side does so also.

Lemma 2.7. *If* $b_n \downarrow 0$, *then* (i) *the series* $\sum b_n \sin nx$ *converges, uniformly on each* $(\varepsilon, \pi - \varepsilon)(\varepsilon > 0)$ *to a function* $g(x)$; (ii) *there is a right-hand neighborhood of* 0 *in which* g *and all the partial sums of the series are positive; and* (iii) $g(x) \sin x \in L$ *(so that the* b_n *are generalized Fourier coefficients).*[1]

The interval of positivity depends on the b_n; it may be arbitrarily small.

Proof. By Lemma 2.3, we can write

$$(2 \sin x) \sum_{k=1}^n b_k \sin kx = -b_2(1 - \cos x) + \sum_{k=2}^{n-1} (b_{k-1} - b_{k+1})(1 - \cos kx)$$

$$\text{(2.8)} \qquad\qquad\qquad + b_{n-1}(1 - \cos nx) + b_n(1 - \cos(n+1)x).$$

Since $b_n \to 0$, $b_{k-1} - b_{k+1} \geq 0$, and $\sum(b_{k-1} - b_{k+1})$ converges, the right-hand side of (2.8) approaches a limit as $n \to \infty$, uniformly on $(0, \pi)$. Hence so does the left-hand side, and its limit is bounded. This proves (i) and (iii).

We note that by (2.8)

$$2 \sum_{k=1}^n b_k \sin kx \geq -b_2 \frac{1 - \cos x}{\sin x} = -b_2 \tan \tfrac{1}{2}x,$$

so that the partial sums of $\sum b_k \sin kx$ are bounded below outside a neighborhood of π. This simple remark can replace (ii) in many applications of Lemma 2.7. In addition, ordinary partial summation yields

$$\sum_{k=1}^n b_k \sin kx = \sum_{k=1}^{n-1} \tilde{D}_k(x)(b_k - b_{k+1}) + \tilde{D}_n(x) b_n,$$

---

[1] HARTMAN and WINTNER [1]; ALJANČIĆ, BOJANIĆ and TOMIĆ [2]; ZYGMUND [2], vol. 1, p. 229; PAK [1], [2].

where $\tilde{D}_n$ is the conjugate Dirichlet kernel,

$$\tilde{D}_n(x) = \sum_{k=1}^{n} \sin \ kx = \frac{\sin \frac{1}{2}(n+1)x \sin \frac{1}{2}nx}{\sin \frac{1}{2}x}.$$

Since $\tilde{D}_n(x)$ is uniformly bounded outside a neighborhood of 0, we see that in fact the partial sums of $\sum b_k \sin \ kx$ are bounded below on the whole interval $(0, \pi)$.

Letting $n \to \infty$ in (2.8), we have

(2.9) $\quad (2 \sin x) g(x) = -b_2(1 - \cos x) + \sum_{k=2}^{\infty} (b_{k-1} - b_{k+1})(1 - \cos kx).$

Since all the terms on the right, except the first, are positive, we have

$$4 \liminf_{x \to 0+} x^{-1} g(x) \geq -b_2 + \sum_{k=2}^{\infty} k^2(b_{k-1} - b_{k+1}),$$

whether the series on the right converges or diverges. If it diverges, we have $g(x)/x \to +\infty$ as $x \to 0+$, and in particular $g(x) \geq 0$ in a right-hand neighborhood of 0. By Lemma 2.6, the series converges or diverges with $\sum k \, b_k$. If both series converge, $\sum b_k(\sin \ kx/\sin x)$ converges uniformly since $|\sin \ kx| \leq k \sin x$; then

$$\lim_{x \to 0+} x^{-1} g(x) = \sum_{k=1}^{\infty} k \, b_k,$$

and again $g(x) \geq 0$ in a right-hand neighborhood of 0.

To show that the partial sums $s_n(x)$ of $\sum b_n \sin \ nx$ have a common interval of positivity, suppose first that $\sum k^2(b_{k-1} - b_{k+1})$ (or equivalently $\sum k \, b_k$) converges. Since $\sum k \, b_k$ converges and $b_k \downarrow$,

$$\sum_{n/2}^{\infty} k \, b_k \geq \sum_{n/2}^{n} k \, b_k \geq b_n \sum_{n/2}^{n} k,$$

and so $n^2 \, b_n \to 0$.

We have, by Lemma 2.3,

(2.10) $\quad \dfrac{2}{x} \sum_{k=n}^{\infty} b_k \sin \ kx = -b_n \dfrac{1 - \cos(n-1)x}{x \sin x} - b_{n+1} \dfrac{1 - \cos nx}{x \sin x}$

$$+ \sum_{k=n+1}^{\infty} k^2(b_{k-1} - b_{k+1}) \frac{1 - \cos kx}{k^2 \, x \sin x},$$

and since $1 - \cos \ kx \leq k^2 x^2$ the left-hand side of (2.10) approaches zero as $n \to \infty$, uniformly in x. Unless all $b_n = 0$, we already know that $g(x)/x \to \sum k \, b_k > 0$ as $x \to 0+$. Hence $x^{-1} s_n(x)$ converges uniformly to a positive limit in some interval $0 \leq x \leq \delta$. Since $x^{-1} s_n(x)$ has a strictly positive limit (or else is identically 0) as $x \to 0+$, (ii) of Lemma 2.7 is established if $\sum k^2(b_{k-1} - b_{k+1})$ converges.

If, on the other hand, $\sum k^2(b_{k-1} - b_{k+1})$ diverges, (2.8) yields, for $n > m$,

$$\frac{2}{x} \sum_{k=1}^{n} b_k \sin \ kx \geq \frac{x}{\sin x} \left\{ -b_2 \frac{1 - \cos x}{x^2} + \sum_{k=2}^{m} k^2(b_{k-1} - b_{k+1}) \frac{1 - \cos kx}{k^2 \, x^2} \right\}.$$

If $x < \pi/(2\,m)$, we have

$$\frac{1-\cos kx}{k^2 x^2} \geq \frac{1}{\pi}, \quad 2 \leq k \leq m;$$

$$\frac{1-\cos x}{x^2} \leq \frac{1}{2},$$

so

$$\frac{2\,s_n(x)}{x} \geq \frac{x}{\sin x}\left\{-\frac{1}{2}b_2 + \frac{1}{\pi}\sum_{k=2}^{m} k^2(b_{k-1}-b_{k+1})\right\}, \quad 0 < x < \pi/(2\,m).$$

We have only to take m so large that the expression in $\{\ldots\}$ is positive.

Remark. In proving Lemma 2.7 we did not use the full force of the hypothesis $b_n \downarrow 0$, but only that $b_n \to 0$ and $b_n - b_{n+2} \geq 0$. Hence the conclusions hold, for example, for series $\sum c_n \sin(2n+1)x$ with $c_n \downarrow 0$ (cf. FEJÉR [1]). This suggests the following question.

Question 2.11. For what sequences $\{n_k\}$ of integers is $\sum c_k \sin n_k x \geq 0$ (or bounded below) in a right-hand neighborhood of 0 provided that $c_k \downarrow 0$?

Lemma 2.12[1]). *If $a_n \downarrow 0$ the series $\sum a_n \cos nx$ converges, uniformly on each $(\varepsilon, \pi - \varepsilon)(\varepsilon > 0)$ to a function $f(x)$.*

This can be proved like (i) of Lemma 2.7.

If $b_n \downarrow 0$ and $\{b_n\}$ is also convex, it follows by ordinary partial summation, and the positivity of the conjugate Fejér kernel, that $g(x) \geq 0$ on $(0, \pi)$. For special convex sequences, even more can be established.

Lemma 2.13. *If $0 < \gamma < 1$, the series $\sum n^{\gamma-1} \sin nx$ has a (positive) sum which is asymptotic to $\Gamma(\gamma)\,\sin\frac{1}{2}\pi\gamma\,x^{-\gamma}$ as $x \to 0+$; its partial sums are bounded below, and uniformly $O(x^{-\gamma})$.*

The conjugate of the series in 2.13 has similar properties.

Lemma 2.14. *If $0 < \gamma < 1$, the series $\sum n^{\gamma-1} \cos nx$ has a (positive) sum which is asymptotic to $\Gamma(\gamma)\,\cos\frac{1}{2}\pi\gamma\,x^{-\gamma}$ as $x \to 0$; its partial sums are uniformly $O(x^{-\gamma})$.*

The partial sums of $\sum n^{\gamma-1}\cos nx$ are bounded below if and only if $\gamma \leq \gamma_0$, where[2]) γ_0 is about 0.6915562.

We prove Lemma 2.13; Lemma 2.14 is proved similarly.

Since

$$\int_{k-\frac{1}{2}}^{k+\frac{1}{2}} \sin xt\,dt = \frac{2\sin\frac{1}{2}x}{x}\,\sin kx,$$

---

[1]) The results stated in Lemmas 2.12, 2.13 and 2.14 can be collected from ZYGMUND [1], chapter 5.

[2]) For the existence of γ_0 see ZYGMUND [2], vol. 1, p. 191; for the numerical value see CHURCH [1]; LUKE, FAIR, COOMBS and MORAN [1].

we have

$$s_n(x) = \frac{x}{2 \sin \frac{1}{2}x} \sum_{k=1}^{n} k^{\gamma-1} \int_{k-\frac{1}{2}}^{k+\frac{1}{2}} \sin xt\, dt,$$

(2.15)
$$2\, x^{-1} \sin \tfrac{1}{2}x\, s_n(x) - \int_{\frac{1}{2}}^{n+\frac{1}{2}} t^{\gamma-1} \sin xt\, dt$$

$$= \sum_{k=1}^{n} \int_{k-\frac{1}{2}}^{k+\frac{1}{2}} (k^{\gamma-1} - t^{\gamma-1}) \sin xt\, dt.$$

Since $|k^{\gamma-1} - t^{\gamma-1}| \le C\, t^{\gamma-2}$ for $k - \frac{1}{2} \le t \le k + \frac{1}{2}$, the absolute value of the right-hand side of (2.15) does not exceed

$$C \int_0^\infty t^{\gamma-2} |\sin\, xt|\, dt = C\, x^{1-\gamma} \int_0^\infty u^{\gamma-2} |\sin u|\, du,$$

and so, in the first place, the left-hand side of (2.15) is bounded below. Since

$$\int_{\frac{1}{2}}^{n-\frac{1}{2}} t^{\gamma-1} \sin xt\, dt = \int_0^{n-\frac{1}{2}} t^{\gamma-1} \sin xt\, dt - \int_0^{\frac{1}{2}} t^{\gamma-1} \sin xt\, dt,$$

and the first term on the right is positive and the second is bounded, it follows that $s_n(x)$ is uniformly bounded below. Furthermore,

$$\int_{\frac{1}{2}}^{n-\frac{1}{2}} t^{\gamma-1} \sin xt\, dt = x^{-\gamma} \left\{ \int_0^\infty u^{\gamma-1} \sin u\, du - \int_0^{\frac{1}{2}x} u^{\gamma-1} \sin u\, du - \int_{(n-\frac{1}{2})x}^\infty u^{\gamma-1} \sin u\, du \right\}.$$

As $x \to 0+$, the second term on the right is $O(x^{\gamma+1})$, the third is $O(1)$, and so $s_n(x) = O(x^{-\gamma})$, uniformly in n. Letting $n \to \infty$, we have

$$\int_{\frac{1}{2}}^\infty t^{\gamma-1} \sin xt\, dt = x^{-\gamma} \left\{ \int_0^\infty u^{\gamma-1} \sin u\, du + O(x^{\gamma+1}) \right\},$$

and so

$$\sum n^{\gamma-1} \sin nx \sim x^{-\gamma} \int_0^\infty u^{\gamma-1} \sin u\, du = \Gamma(\gamma) \sin \tfrac{1}{2}\pi\gamma\, x^{-\gamma}.$$

Lemma 2.16. *The series* $\sum n^{-1} \sin nx$ *has uniformly bounded partial sums, positive on* $(0, \pi)$.

Proof. We proceed by induction. Let $s_n(x)$ be the partial sums of the series. For $n = 1$ we have $s_n(x) > 0$ for $0 < x < \pi$. Suppose this is true for $n-1$. Since

$$2 \sin \tfrac{1}{2}x\, s_n'(x) = 2 \sin \tfrac{1}{2}x \sum_{k=1}^{n} \cos kx = \sin(n+\tfrac{1}{2})x - \sin \tfrac{1}{2}x,$$

it follows that at a minimum x_0 of $s_n(x)$ we have $\sin(n+\frac{1}{2})x_0 = \sin\frac{1}{2}x_0$, and consequently $\cos(n+\frac{1}{2})x_0 = \pm\cos\frac{1}{2}x_0$. Therefore

$$\sin nx_0 = \sin(n+\tfrac{1}{2})x_0\cos\tfrac{1}{2}x_0 - \cos(n+\tfrac{1}{2})x_0\sin\tfrac{1}{2}x_0$$

$$= \sin\tfrac{1}{2}x_0(\cos\tfrac{1}{2}x_0 \mp \cos\tfrac{1}{2}x_0) \geq 0.$$

Hence

$$s_n(x_0) = s_{n-1}(x_0) + n^{-1}\sin nx_0 \geq s_{n-1}(x_0) > 0.$$

That is, the minimum of $s_n(x)$ is positive[1]).

On the other hand,

$$s_n(x) = \sum_{k=1}^{1/x} + \sum_{k=1/x}^{n} \frac{\sin kx}{k}.$$

Since $|\sin kx| \leq kx$,

$$|s_n(x)| \leq \sum_{k=1}^{1/x} \frac{kx}{k} + x \sup_{1/x<p<m\leq n}\left|\sum_{k=p}^{m}\sin kx\right|,$$

where the second sum has been estimated by the second mean-value theorem. Therefore

$$|s_n(x)| \leq 1 + x \sup_{1/x<p\leq m\leq n}\left|\frac{\cos(p+\frac{1}{2})x - \cos(m+\frac{1}{2})x}{2\sin\frac{1}{2}x}\right|$$

$$\leq 1 + \pi,$$

since $\sin\frac{1}{2}x \geq x/\pi$.

Lemma 2.17. *If $-1 < \gamma < 0$ the series $\sum n^{\gamma-1}\sin nx$ has positive partial sums, and its sum is bounded below by $C_\gamma x^{-\gamma}$. For $\gamma = -1$, its sum is asymptotic to $C\,x\log(1/x)$.*

Since $\gamma < 0$, ordinary partial summation together with Lemma 2.16 shows that the partial sums are positive. Since

$$\sum n^\gamma \cos nx \geq C_\gamma x^{-\gamma-1}, \qquad -1 < \gamma < 0,$$

by Lemma 2.14, and the series on the left is a Fourier series, we can integrate this inequality to get

$$\sum n^{\gamma-1}\sin nx \geq C_\gamma x^{-\gamma}.$$

For $\gamma = -1$ the series is an integral of

$$\sum n^{-1}\cos nx = \log(\tfrac{1}{2}\csc\tfrac{1}{2}x) \sim \log(1/x).$$

---

[1]) This proof was given by LANDAU; for other proofs, and references, see HYLTÉN-CAVALLIUS [1], TURÁN [1], [2], and TOMIĆ [1], [2].

§ 3. Theorems with positive or decreasing functions

Theorem 3.1[1]). *Sines,* $-1 \leq \gamma < 0$. *Let* $g(x) \geq 0$ *in a right-hand neighborhood of* 0, *with* $x\, g(x) \in L$. *Let* b_n *be the generalized sine coefficients of* g. *If* $-1 < \gamma < 0$, *then* $\sum n^{\gamma-1}|b_n|$ *converges if and only if* $x^{-\gamma} g(x) \in L$. *If* $\gamma = -1$, $\sum n^{-2}|b_n|$ *converges if and only if* $x \log(1/x)\, g(x) \in L$.

Theorem 3.2[1]). *Cosines,* $-2 < \gamma < 0$. *Let* $f(x) \geq 0$ *in a neighborhood of* 0, *with* $x^2 f(x) \in L$. *Let* a_n *be the generalized cosine coefficients of* f. *Then* $\sum n^{\gamma-1}|a_n|$ *converges if and only if* $x^{-\gamma} f(x) \in L$.

Question 3.3. Generalize Theorems 3.1 and 3.2 to smaller values of γ.

Theorem 3.4[2]). *Sines,* $0 \leq \gamma < 1$. *Let* $g(x) \downarrow$ *in* $(0, \delta)$, $x\, g(x) \in L$, g *bounded on* (δ, π). *Let* b_n *be the generalized sine coefficients of* g. *If* $0 \leq \gamma < 1$, $\sum n^{\gamma-1}|b_n|$ *converges if and only if* $x^{-\gamma} g(x) \in L$.

Some condition on g in the neighborhood of π is essential if $x^{-\gamma} g(x) \in L$ is to imply $\sum n^{\gamma-1}|b_n|$ convergent. This was pointed out to me by S. M. EDMONDS and I reproduce her example. If g is negative and decreasing in $(0, \pi)$, then $-g(\pi - x)$ is positive and decreasing and its generalized sine coefficients are $(-1)^n b_n$, so $\sum n^{\gamma-1}|b_n|$ converges if and only if $x^{-\gamma} g(\pi - x) \in L$, i.e. $(\pi - x)^{-\gamma} g(x) \in L$. Consequently we have a counterexample to Theorem 3.4 if $x^{-\gamma} g(x) \in L$ but $(\pi - x)^{-\gamma} g(x)$ is not in L (and consequently g is not bounded below).

Theorem 3.5[3]). *Cosines,* $0 \leq \gamma < 1$. *Let* $f(x) \downarrow$ *in* $(0, \delta)$, $f \in L$, f *bounded below on* (δ, π). *Let* a_n *be the cosine coefficients of* f. *If* $0 < \gamma < 1$, $\sum n^{\gamma-1}|a_n|$ *converges if and only if* $x^{-\gamma} f(x) \in L$. *If* $\gamma = 0$, $\sum n^{-1}|a_n|$ *converges if and only if* $f(x) \log(1/x) \in L$.

Question 3.6. What condition is necessary and sufficient for $f \in L$ under the hypotheses of Theorem 3.5?

Note that if $g(x) \downarrow$ on $(0, \pi)$ we have $b_n \geq 0$. There is therefore no point in extending Theorem 3.4 to larger values of γ, since the theorems of later sections with $b_n \geq 0$ would include such extensions. This is probably true for Theorem 3.5 as well, since the natural extension to larger γ would have f convex and so $a_n \geq 0$ (ZYGMUND [2], vol. 1, p. 183).

If we replace decreasing functions by positive functions, Theorems 3.4 and 3.5 fail, in the $\int \rightarrow \sum$ direction, for $\frac{1}{2} < \gamma < 1$. This was pointed out to me by R. ASKEY,

---

[1]) BOAS [4].
[2]) ZYGMUND [1] for $\gamma = 0$; SZ.-NAGY [1] for $0 \leq \gamma < 1$; also EDMONDS [1], part III, where a more general result is given (cf. § 8 below).
[3]) ZYGMUND [1] for $\gamma = 0$; SZ.-NAGY [1] for $0 \leq \gamma < 1$.

with the examples

$$C + \sum_{n=2}^{\infty} \frac{\cos cn \log n \cos nx}{n^{1/2}(\log n)^2}, \quad C + \sum_{n=2}^{\infty} \frac{\sin cn \log n \sin nx}{n^{1/2}(\log n)^2}, \quad c > 0,$$

with suitable C. Somewhat simpler examples are given by $g(x) = C + g_1(x)$, $f(x) = C + f_1(x)$, where

$$g_1(x) = \sum_{n=2}^{\infty} \pm \frac{\sin nx}{n^{1/2}(\log n)^2}, \quad f_1(x) = \sum_{n=2}^{\infty} \pm \frac{\cos nx}{n^{1/2}(\log n)^2},$$

with the signs chosen so that the series converge uniformly (ZYGMUND [2], vol. 1, p. 219). In fact, $f(x) = C + \sum a_n \cos nx$, $g(x) = \sum b_n \sin nx$, with $x^{-\gamma} f(x)$, $x^{-\gamma} g(x) \in L$ if $0 < \gamma < 1$; but $\sum n^{\gamma-1} |a_n|$ and $\sum n^{\gamma-1} |b_n|$ diverge if $\gamma > \frac{1}{2}$.

Question 3.7. Find examples to show that Theorems 3.4 and 3.5 fail for positive functions when $0 \le \gamma \le \frac{1}{2}$, or fail in the $\sum \to \int$ direction (or else prove them in these cases).

We may ask for necessary and sufficient conditions for $x^{-\gamma} g(x) \in L$ when $g(x) \ge 0$ or for $x^{-\gamma} f(x) \in L$ $(\gamma > 0)$ or $f(x) \log(1/x) \in L$ when $f(x) \ge 0$. These are provided by the following theorems[1]).

Theorem 3.8. *If* $x g(x) \in L, g(x) \ge 0$ *in* $(0, \delta), g$ *is bounded in* $(\delta, \pi),$ *and* $0 \le \gamma < 1,$ *then* $x^{-\gamma} g(x) \in L$ *if and only if* $\sum n^{\gamma-1} b_n$ *converges (not necessarily absolutely).*

Theorem 3.9. *If* $f(x) \in L,$ $f(x) \ge 0$ *in* $(0, \delta),$ *and* $0 < \gamma < 1,$ *then* $x^{-\gamma} f(x) \in L$ *if and only if* $\sum n^{\gamma-1} a_n$ *converges;* $f(x) \log(1/x) \in L$ *if and only if* $\sum n^{-1} a_n$ *converges.*

Question 3.10. What conditions are necessary and sufficient for $\sum n^{\gamma-1} |b_n|$ or $\sum n^{\gamma-1} |a_n|$ to converge, $0 \le \gamma < 1$, provided $g(x) \ge 0$ or $f(x) \ge 0$?

Proof of Theorem 3.1, $\sum \to \int$. We prove a result that is more general in two ways: we replace $g(x) dx$ by $d G(x)$, and we do not assume absolute convergence of the series.

3.11. *Let* $G(x) \downarrow$ *on some* $(0, \varepsilon), \varepsilon > 0,$ *and let* G *be of bounded variation on* $(\varepsilon, \pi),$ *with* $\int x \, dG(x)$ *finite. Let*

$$(3.12) \qquad b_n = -2 \pi^{-1} \int \sin nx \, dG(x).$$

Then if $\sum n^{\gamma-1} b_n$ *converges and* $-1 < \gamma < 0$ *it follows that* $\int x^{-\gamma} dG(x)$ *is finite. If* $\gamma = -1,$ $\int x \log(1/x) d G(x)$ *is finite.*

By Lemma 2.17, $\sum n^{\gamma-1} \sin nx$ is uniformly convergent on $(0, \varepsilon)$, with positive partial sums. By Fatou's lemma,

$$-\int_0^{\varepsilon} (\sum n^{\gamma-1} \sin nx) \, dG(x) \le -\sum n^{\gamma-1} \int_0^{\varepsilon} \sin nx \, dG(x)$$

$$= -\sum n^{\gamma-1} \int_0^{\pi} \sin nx \, dG(x) + \sum n^{\gamma-1} \int_{\varepsilon}^{\pi} \sin nx \, dG(x).$$

---

[1]) BOAS [4]; Theorem 3.8 was given by EDMONDS [2] in a more general form (Theorem 8.7 below).

The first series on the right is $\frac{1}{2}\pi \sum n^{\gamma-1} b_n$; the second is absolutely convergent since $\gamma < 0$. Hence

$$- \int_0^\varepsilon (\sum n^{\gamma-1} \sin nx \, dG(x) < \infty.$$

The sum of the series exceeds a multiple of $x^{-\gamma}$ if $\gamma > -1$ (Lemma 2.17), and hence

$$- \int_0^\varepsilon x^{-\gamma} \, dG(x) < \infty.$$

The conclusion for $\gamma = -1$ follows in the same way.

Proof of Theorem 3.1, $\int \to \sum$. Here again we prove a more general proposition.

3.13. *If G is of bounded variation on every (ε, π), $\varepsilon > 0$, if b_n are defined by (3.12), and if $\int x^{-\gamma} |d\,G(x)|$ is finite, then $\sum n^{\gamma-1}|b_n|$ converges if $-1 < \gamma < 0$; if $\int x^{-1} \log(1/x)| dG(x)|$ is finite then $\sum n^{-2}|b_n|$ converges.*

We have, if $-1 < \gamma < 0$,

$$\tfrac{1}{2}\pi \sum n^{\gamma-1}|b_n| \le \sum n^{\gamma-1} \int |\sin nx| \, |dG(x)|$$
$$\le \int |d\,G(x)| \sum n^{\gamma-1}|\sin nx|.$$

Since

$$\sum n^{\gamma-1}|\sin nx| \le \sum_1^{1/x} + \sum_{1/x}^\infty \le x \sum_1^{1/x} n^\gamma + \sum_{1/x}^\infty n^{\gamma-1} = O(x^{-\gamma}),$$

the result follows for $-1 < \gamma < 0$; the proofs for $\gamma = -1$ and for Theorem 3.2 are similar.

The generalized form of the $\sum \to \int$ part of Theorem 3.2, corresponding to 3.11, reads as follows.

3.14. *Let $F(x) \downarrow$ on some $(0, \varepsilon)$, $\varepsilon > 0$, and let F be of bounded variation on (ε, π), with*

$$- \int_0^\varepsilon x^2 dF(x) < \infty.$$

Let

$$a_n = 2\pi^{-1} \int (1 - \cos nx) \, dF(x).$$

Then if $\sum n^{\gamma-1} a_n$ converges and $-2 < \gamma < 0$ it follows that $\int x^{-\gamma}|d\,F(x)| < \infty$.

The proof follows the same lines as that of 3.11; the appeal to Lemma 2.17 at the end of the $\sum \to \int$ part of the proof of Theorem 3.1 is replaced by

$$\sum n^{\gamma-1}(1 - \cos nx) \ge C x^2 \sum_{n=1}^{1/x} n^{\gamma+1} \ge C x^{-\gamma}, \quad -2 < \gamma < 0.$$

This completes our discussion of the case $\gamma < 0$. We turn now to Theorems 3.4 and 3.5, where $0 \le \gamma < 1$. Again we can prove somewhat more than the symmetric statements of the theorems.

Proof of Theorem 3.4, $\sum \to \int$. We prove more by proving the corresponding part of Theorem 3.8 instead.

First consider $\gamma = 0$. The partial sums of $\sum n^{-1} \sin nx$ are positive (Lemma 2.16); hence, as in the corresponding part of Theorem 3.1,

$$\int_0^\varepsilon \left(\sum n^{-1} \sin nx\right) g(x)\, dx \le \sum n^{-1} \int g(x) \sin nx\, dx - \sum n^{-1} \int_\varepsilon^\pi g(x) \sin nx\, dx.$$

The second series on the right converges since it is of the form $\sum n^{-1} \beta_n$ with β_n the Fourier sine coefficients of an integrable function (cf. 4.11 below), and so

$$\int_0^\varepsilon \left(\sum n^{-1} \sin nx\right) g(x)\, dx < \infty.$$

Since the sum of the series $\left(\text{namely } \tfrac{1}{2}(\pi - x)\right)$ is bounded away from 0,

$$\int_0^\varepsilon g(x)\, dx < \infty$$

and so $g \in L$.

Next consider $0 < \gamma < 1$. Since $\sum n^{\gamma - 1} b_n$ converges, $\sum n^{-1} b_n$ converges, so $g \in L$ by what has just been proved. Since $\sum n^{\gamma - 1} \sin nx$ has its partial sums bounded below (Lemma 2.13), say by $-K$, and $g \in L$, we have by Fatou's lemma

$$\int_0^\delta \left(\sum n^{\gamma - 1} \sin nx + K\right) g(x)\, dx \le \sum n^{\gamma - 1} \int_0^\delta g(x) \sin nx\, dx + K \int_0^\delta g(x)\, dx.$$

The second term on the right is finite. The first is

$$\tfrac{1}{2} \pi \sum n^{\gamma - 1} b_n - \sum n^{\gamma - 1} \int_\delta^\pi g(x) \sin nx\, dx$$

$$= \tfrac{1}{2} \pi \sum n^{\gamma - 1} b_n - \lim_{N \to \infty} \int_\delta^\pi g(x) \sum_{n=1}^N n^{\gamma - 1} \sin nx\, dx,$$

and this is finite since $\sum n^{\gamma - 1} \sin nx$ converges uniformly on (δ, π) and $g \in L$. Hence $g(x) \sum n^{\gamma - 1} \sin nx \in L(0, \delta)$. By Lemma 2.13, $g(x) x^{-\gamma}$ belongs to $L(0, \delta)$ and hence to $L(0, \pi)$.

Proof of Theorem 3.4, $\int \to \sum$. We deduce this from 3.14 by integration by parts; this is the only place where we actually use the assumption that $g(x)$ decreases. We give the proof when g is of bounded variation in each (δ, π), $\delta > 0$. This shows the essential idea; to remove this extra condition is a routine matter.

For $n > 0$ we have

$$\tfrac{1}{2} \pi b_n = \int g(x) \sin nx\, dx = \frac{1 - \cos nx}{n} g(x) \Big|_0^\pi + n^{-1} \int (1 - \cos nx)\, dg(x).$$

Since g decreases on $(0, \delta)$ and $x^{-\gamma} g(x) \in L$, we have $g(x) = o(x^{\gamma - 1})$; in fact,

$$g(x) x^{1 - \gamma}/(1 - \gamma) = g(x) \int_0^x t^{-\gamma}\, dt \le \int_0^x t^{-\gamma} g(t)\, dt = o(1).$$

Hence $(1 - \cos nx) g(x) \to 0$ as $x \to 0$. We are assuming $g(\pi)$ finite, so $\sum n^{\gamma - 2} g(\pi)$ converges. Hence $\sum |b_n| n^{\gamma - 1}$ converges if

$$\sum n^{\gamma - 2} \left| \int (1 - \cos nx)\, dg(x) \right|$$

converges. We now apply 3.14 with exponent $\gamma-1$. To do this, we have to know that $\int x^{1-\gamma}|dg(x)| < \infty$. For (δ, π) this is obvious. For $(0, \delta)$, it is

$$\int_0^{\delta} x^{1-\gamma}\, dg(x) < \infty;$$

since $g(x) = o(x^{\gamma-1})$, this is equivalent (by integration by parts) to

$$\left| \int_0^{\delta} x^{-\gamma}\, g(x)\, dx \right| < \infty.$$

Proof of Theorem 3.5, $\sum \to \int$. Again we prove more by proving the corresponding part of Theorem 3.9. The proof runs parallel to the corresponding part of Theorem 3.3 except that $\sum n^{\gamma-1} \cos nx$ does not have its partial sums bounded below for all γ in $0 < \gamma < 1$. However, the sum of the series is bounded below (Lemma 2.14), and therefore so are its Abel means. Hence if $f(x) \geq 0$ on $(0, \delta)$

(3.15) $$\int_0^{\delta} f(x) \sum n^{\gamma-1} \cos nx\, dx \leq \liminf_{r \to 1} \sum r^n\, n^{\gamma-1} \int_0^{\delta} f(x) \cos nx\, dx$$

$$= \liminf_{r \to 1} \sum r^n\, n^{\gamma-1} \left\{ \tfrac{1}{2} \pi\, a_n - \int_{\delta}^{\pi} f(x) \cos nx\, dx \right\}.$$

Since f is integrable on (δ, π) and $\sum n^{\gamma-1} \cos nx$ is bounded there, Parseval's theorem holds for f, restricted to (δ, π), and $\sum n^{\gamma-1} \cos nx$, with Abel summability. This means in particular that

$$\sum r^n\, n^{\gamma-1} \int_{\delta}^{\pi} f(x) \cos nx\, dx$$

approaches a limit as $r \to 1$. Since $\sum a_n\, n^{\gamma-1}$ converges, $\sum r^n\, n^{\gamma-1}\, a_n$ also approaches a limit. Hence the left-hand side of (3.15) is finite. It then follows (Lemma 2.14) that $x^{-\gamma} f(x) \in L$.

The same proof works when $\gamma = 0$. In this case we could alternatively appeal to a theorem of HARDY and LITTLEWOOD (see 5.17), which states that $\sum n^{-1} a_n$ converges if and only if

$$\int_{-\delta}^{\delta} x^{-1}\, dx \int_{-x}^{x} f(t)\, dt$$

converges; this condition implies $f(x) \log(1/x) \in L$ when $f(x) \geq 0$ on $(0, \delta)$.

Finally we prove the $\int \to \sum$ parts of Theorems 3.8 and 3.9. For Theorem 3.8, we have

(3.16) $$\sum_m^M n^{\gamma-1} b_n = 2\pi^{-1} \sum_m^M n^{\gamma-1} \int g(t) \sin nt\, dt$$

$$= 2\pi^{-1} \int g(t) \sum_m^M n^{\gamma-1} \sin nt\, dt.$$

The sums in the last integral are uniformly $O(t^{-\gamma})$ by Lemmas 2.13 and 2.16. Since $x^{-\gamma} g(x) \in L$, we have dominated convergence on the right of (3.16), and so the left-hand side of (3.16) approaches zero as m and $M \to \infty$. The argument for Theorem 3.9 is similar. (Note that this part of the proof makes no use of the positivity of g or f.)

§ 4. Theorems with positive or decreasing coefficients

We continue to arrange the theorems in order of increasing γ.

Theorem 4.1[1]). *Sines,* $0 \le \gamma \le 1$. *Let* $b_n \downarrow$ *ultimately to* 0. *Then* $\sum b_n \sin nx$ *converges to* $g(x)$, *and* $x^{-\gamma} g(x) \in L$ *if and only if* $\sum n^{\gamma-1} b_n$ *converges.*

Theorem 4.2[2]). *Cosines,* $0 < \gamma < 1$. *Let* $a_n \downarrow$ *ultimately to* 0. *Then* $\sum a_n \cos nx$ *converges to* $f(x)$, *and* $x^{-\gamma} f(x) \in L$ *if and only if* $\sum n^{\gamma-1} a_n$ *converges.*

Note that (by 4.11, 4.12) the $\int \to \sum$ parts of these theorems hold if[3]) $x^{-\gamma} \varphi(x) \in L$ without any assumptions about λ_n except that they are Fourier coefficients.

Theorems 4.1 and 4.2 fail in a trivial way if $\{b_n\}$ or $\{a_n\}$ is assumed positive instead of monotonic, since $\sum b_n \sin nx$ or $\sum a_n \cos nx$ might not converge. (But see § 8 for a discussion of intermediate cases.) They presumably fail even if b_n or a_n are assumed to be positive Fourier coefficients. There are examples to show this when $0 < \gamma < \frac{1}{2}$, and when $\gamma = 1$, but I do not know any when $\frac{1}{2} < \gamma < 1$. For $\gamma = 1$, $\sum \to \int$ fails since $g(x) = \sum n^{-3/2} \sin 2^n x$ has $\sum b_n < \infty$ but $x^{-1} g(x)$ is not integrable[4]). (See Theorem 5.27.) For $0 < \gamma < \frac{1}{2}$, $\sum \to \int$ also fails. The following construction was suggested by R. ASKEY. If $0 < \alpha < 1$ and $\frac{1}{2}\alpha < \beta < 1 - \frac{1}{2}\alpha$ the series[5])

$$\sum n^{-\beta} \exp(i\, n^{\alpha}) \cos nx$$

is a Fourier series and its sum $F(x)$ has the form

$$C\, x^{-(1-\beta-\alpha/2)/(1-\alpha)} \exp(i\, \delta\, x^{-1/(1-\alpha)}) \{1 + O(x^{\alpha/(1-\alpha)})\},$$

where δ is real and not zero. Consequently if $\beta = \gamma + \alpha(\frac{1}{2} - \gamma)$ (a value which satisfies $\frac{1}{2}\alpha < \beta < 1 - \frac{1}{2}\alpha$) we have $\int |F(x)|\, x^{-\gamma}\, dx = \infty$. The same is therefore true for either the real part or the imaginary part of F, i. e. for one of

$$\sum n^{-\beta} \cos n^{\alpha} \cos nx, \quad \sum n^{-\beta} \sin n^{\alpha} \cos nx.$$

Also $\sum n^{-\beta} \cos nx \sim C x^{\beta-1}$ as $x \to 0$, so $x^{-\gamma} \sum n^{-\beta} \cos nx \in L$ if $\beta > \gamma$. Hence one of the functions f defined by the Fourier series (with positive coefficients a_n)

---

[1]) EDMONDS [1], part III, where more general results are given (cf. § 8 below); SUNOUCHI [1], BOAS [2]; for $\gamma = 0$, YOUNG [1].

[2]) SUNOUCHI [1], BOAS [2].

[3]) For the notation see p. 1).

[4]) The first example of this phenomenon was given by O'SHEA [1].

[5]) HARDY [1], WAINGER [1].

3*

$$\sum n^{-\beta}(2+\cos n^{\alpha})\cos nx, \quad \sum n^{-\beta}(2+\sin n^{\alpha})\cos nx$$

does not have $x^{-\gamma}f(x)\in L$. On the other hand $\sum n^{\gamma-1}a_n$ converges since $\beta > \gamma$.

Question 4.3. Find examples for $\frac{1}{2} < \gamma < 1$.

Two questions now arise: if λ_n are positive sine or cosine coefficients, what conditions are necessary and sufficient for $\sum n^{\gamma-1}\lambda_n$ to converge, and what conditions are necessary and sufficient for $x^{-\gamma}\varphi(x)\in L$? The first of these questions is answered by the following theorem.

Theorem 4.4[1]). *Sines or cosines,* $0 < \gamma < 1$. *If* $\lambda_n \geq 0$ *and* λ_n *are the Fourier sine or cosine coefficients of* φ, *and if* $0 < \gamma \leq 1$ *(for sine series) or* $0 < \gamma < 1$ *(for cosine series), then*

(4.5) $$\sum n^{\gamma-1}\lambda_n < \infty$$

if and only if

(4.6) $$\int_{a+}^{\pi} (x-a)^{-\gamma}\varphi(x)\,dx \quad converges, \quad 0 \leq a < \pi.$$

Cf. also Theorem 5.10.

The $\sum \to \int$ part holds with λ_n of arbitrary sign if $\sum n^{\gamma-1}|\lambda_n|$ converges.

Question 4.7. If $\lambda_n \geq 0$ and λ_n are the Fourier sine or cosine coefficients of φ, and $0 < \gamma < 1$, what condition on λ_n is necessary and sufficient for $x^{-\gamma}\varphi(x)\in L$, or for $(x-a)^{-\gamma}\varphi(x)\in L$ $(0 < a < \pi)$?

Theorem 4.8[2]). *Sines,* $1 < \gamma < 2$. *Let* $b_n \geq 0$ *ultimately. Then* $\sum n^{\gamma-1}b_n$ *converges if and only if* b_n *are the Fourier sine coefficients of a continuous function* g *such that either* $x^{-\gamma}g(x)\in L$ *or* $|x-a|^{-\gamma}\{g(x)-g(a)\}\in L$ *for all a,* $0 \leq a < \pi$.

Theorem 4.9[3]). *Cosines,* $1 \leq \gamma < 3$. *Let* $a_n \geq 0$ *ultimately. Then* $\sum n^{\gamma-1}a_n$ *converges* $(\gamma > 1)$ *or* $\sum a_n \log n$ *converges* $(\gamma = 1)$ *if and only if* a_n *are the Fourier cosine coefficients of a continuous function* f *such that for* $0 L a < \pi$ *either* $x^{-\gamma}\{f(x)-f(0)\}\in L(1 \leq \gamma < 3)$ *or* $|x-a|^{-\gamma}\{f(x)-f(a)\}\in L(1 \leq \gamma < 2)$.

ROBERTSON [1] has obtained similar theorems when $\{\lambda_n\}$ can be partitioned into k disjoint sequences each of which ultimately increases or decreases to 0.

It is easiest to begin with the $\sum \to \int$ parts of Theorems 4.8 and 4.9, and the $\int \to \sum$ parts of Theorems 4.1 and 4.2.

Proof of Theorem 4.8, $\sum \to \int$. Assume that $b_n \geq 0$ ultimately and that $\sum n^{\gamma-1}b_n$ converges, where $1 < \gamma < 2$. (We actually need only $\sum n^{\gamma-1}|b_n| < \infty$.) Since $\gamma > 1$, $\sum b_n$ converges and so $\sum b_n \sin nx$ converges uniformly to $g(x)$. By Fatou's lemma,

$$\int |x-a|^{-\gamma}|g(x)-g(a)|\,dx \leq \sum |b_n|\int |x-a|^{-\gamma}|\sin nx-\sin na|\,dx$$

$$\leq 2\sum |b_n|\int |x-a|^{-\gamma}\sin \tfrac{1}{2}n(x-a)|\,dx$$

---

[1]) BOAS [7].
[2]) The first part was discovered by O'SHEA [1].
[3]) The first part was proved by HEYWOOD [1], [3].

$$= 2 \sum |b_n| \int_{-a}^{\pi-a} |t|^{-\gamma} |\sin \tfrac{1}{2}nt| \, dt$$

$$= 2 \sum n^{\gamma-1} |b_n| \int_{-na}^{n(\pi-a)} |u|^{-\gamma} |\sin \tfrac{1}{2}u| \, du$$

$$\le A_\gamma \sum |b_n| \, n^{\gamma-1},$$

with

(4.10) $$A_\gamma = 2 \int_0^\infty |u|^{-\gamma} |\sin \tfrac{1}{2}u| \, du.$$

We require $\gamma < 2$ so that the integral in (4.10) converges at 0, and $\gamma > 1$ so that it converges at ∞.

The proof of the corresponding part of Theorem 4.9 is similar. Assume that $a_n \ge 0$ ultimately and that $\sum n^{\gamma-1} |a_n| < \infty$ ($1 < \gamma < 3$) or that $\sum |a_n| \log n$ converges ($\gamma = 1$). Then

$$\tfrac{1}{2} a_0 + \sum a_n \cos nx$$

converges uniformly to $f(x)$, and hence

$$f(0) - f(x) = \sum a_n (1 - \cos nx).$$

If $1 < \gamma < 3$,

$$\int x^{-\gamma} |f(x) - f(0)| \, dx \le \sum |a_n| \int x^{-\gamma} (1 - \cos nx) \, dx$$

$$= \sum |a_n| \, n^{\gamma-1} \int_0^{n\pi} u^{-\gamma} (1 - \cos u) \, du$$

$$< A_\gamma \sum |a_n| \, n^{\gamma-1}.$$

If $1 < \gamma < 2$,

$$\int |x-a|^{-\gamma} |f(x) - f(a)| \, dx \le \sum |a_n| \int |x-a|^{-\gamma} |\cos nx - \cos na| \, dx$$

$$\le 2 \sum |a_n| \int |x-a|^{-\gamma} |\sin \tfrac{1}{2}n(x-a)| \, dx$$

$$= 2 \sum |a_n| \int_{-a}^{\pi-a} |t|^{-\gamma} |\sin \tfrac{1}{2}nt| \, dt$$

$$\le 2 \sum |a_n| \, n^{\gamma-1} \int_{-na}^{n(\pi-a)} |u|^{-\gamma} |\sin \tfrac{1}{2}u| \, du$$

$$\le A_\gamma \sum |a_n| \, n^{\gamma-1}, \quad 1 < \gamma < 2;$$

for $\gamma = 1$, the last integral is $O(\log n)$, instead of $O(n^{\gamma-1})$.

Proof of Theorem 4.1, $\int \to \sum$. We prove considerably more:

4.11. *If* $0 \le \gamma \le 1$ *and* $x^{-\gamma} g(x) \in L$ *(hence* $g \in L$*), and* b_n *are the Fourier sine coefficients of* g, *then* $\sum n^{\gamma-1} b_n$ *converges.*

This is, of course, classical[1]) for $\gamma = 0$, and not appreciably harder in the general case.

The partial sums of $\sum n^{\gamma-1} \sin nx$ are uniformly $O(x^{-\gamma})$ for each γ, $0 \le \gamma \le 1$ (Lemma 2.13 for $0 < \gamma < 1$; 2.16 for $\gamma = 0$; the explicit form of the conjugate Dirichlet kernel for $\gamma = 1$). Hence when $x^{-\gamma} g(x) \in L$,

$$g(x) \sum_{k=1}^{n} k^{\gamma-1} \sin kx$$

is dominated by a constant multiple of $x^{-\gamma}|g(x)|$. By dominated convergence,

$$\tfrac{1}{2}\pi \left| \sum_{p}^{q} n^{\gamma-1} b_n \right| = \left| \int_{p}^{q} n^{\gamma-1} \int_{0}^{\pi} g(x) \sin nx \, dx \right|$$

$$= \left| \int_{0}^{\pi} \left\{ \sum_{p}^{q} n^{\gamma-1} \sin nx \right\} g(x) \, dx \right|$$

$$\to 0 \qquad\qquad (p, q \to \infty).$$

Proof of Theorem 4.2, $\int \to \sum$. Again we can prove more:

4.12. *If* $0 < \gamma < 1$ *and* $x^{-\gamma} f(x) \in L$ *(hence* $f \in L$*), then* $\sum n^{\gamma-1} a_n$ *converges.*
The proof of 4.12 is just like the proof of 4.11.

By combining 4.11, 4.12 with 4.4, we get the following semi-symmetric theorem.

Theorem 4.13[2]). *If* φ *is integrable and* λ_n *are its sine or cosine coefficients, and* $0 < \gamma < 1$*, then* $\sum |\lambda_n| n^{\gamma-1} < \infty$ *implies*

$$\int_{0+}^{\pi} x^{-\gamma} \varphi(x) \, dx$$

convergent; $\int x^{-\gamma}|\varphi(x)| \, dx < \infty$ *implies* $\sum \lambda_n n^{\gamma-1}$ *convergent.*
Cf. also Theorems 8.10 and 8.11.

Proof of Theorem 4.1, $\sum \to \int$. We deduce this by partial summation from the corresponding part of Theorem 4.9.

We assume that $b_n \downarrow 0$ and that $\sum n^{\gamma-1} b_n$ converges. By Lemma 2.6, with $x_k = b_k$ and $y_k = k^{\gamma}$ ($0 < \gamma \le 1$), or $y_k = \log k$ ($\gamma = 0$), we have

$$\sum (b_{n-1} - b_{n+1}) n^{\gamma} \text{ convergent } (0 < \gamma \le 1),$$

$$\sum (b_{n-1} - b_{n+1}) \log n \text{ convergent } (\gamma = 0).$$

In either case, $\sum (b_{n-1} - b_{n+1})$ converges, its terms are ultimately positive, and

$$-b_1 - b_2 + \sum_{2}^{\infty} (b_{n-1} - b_{n+1}) = 0.$$

Applying Lemma 2.3, we have

$$-(2 \sin x) g(x) = -b_1 + \sum (b_{n-1} - b_{n+1}) \cos nx.$$

---

[1]) See, e. g., ZYGMUND [2], vol. 1, p. 59; BARI [1], p. 123; HARDY and ROGOSINSKI [1], p. 31.
[2]) IZUMI and SATÔ [1].

Theorem 4.9, with index $\gamma + 1$, shows that

$$x^{-\gamma-1}(2 \sin x)\, g(x) \in L,$$

whence $x^{-\gamma} g(x) \in L$.

We have not used the full force of the hypotheses: it is enough that $\{b_{2n}\}$ and $\{b_{2n+1}\}$ should both be ultimately decreasing; hence Theorem 4.1 holds, for example, for $\sum c_n \sin (2n+1)\, x$ with $c_n \downarrow 0$. In addition, the only use we made of the decreasing character of $\{b_n\}$ was to get, from the convergence of $\sum n^{\gamma-1} b_n$, the absolute convergence of $\sum (b_{n-1} - b_{n+1})\, n^{\gamma}(\gamma > 0)$ or of $\sum (b_{n-1} - b_{n+1}) \log n$ $(\gamma = 0)$. Hence we could have taken the absolute convergence of the appropriate one of these series as hypothesis, or equally well any condition on $\{b_n\}$ that implies this; for example, it is enough to have $\{b_n\}$ quasimonotonic[1]. For further discussion of the cases $\gamma = 0$, $\gamma = 1$, see § 5.

The proof of Theorem 4.2, $\sum \to \int$, goes in exactly the same way, and the same comments apply.

Question 4.14. In Theorem 4.1, can $b_n \downarrow 0$ and $\sum n^{\gamma-1} b_n < \infty$ be replaced by $b_n \to 0$, $\sum |\Delta b_n| < \infty$ (i.e., $\{b_n\}$ of bounded variation), and $\sum n^{\gamma-1}|b_n| < \infty$?

To get the $\sum \to \int$ parts of Theorems 4.1 and 4.2, we worked down from Theorems 4.9 and 4.8. Now we work up from the $\int \to \sum$ parts of Theorems 4.2 and 4.1 to get the corresponding parts of Theorems 4.8 and 4.9.

Proof of Theorem 4.8, $\int \to \sum$. We suppose that $x^{-\gamma} g(x) \in L$, where $1 < \gamma < 2$, that b_n are the Fourier sine coefficients of g, and that $b_n \geq 0$ ultimately.

Since $\gamma > 1$, $x^{-1} g(x) \in L$ and Theorem 4.1 (with index 1) shows that $\sum b_n$ converges. By Lemma 2.5 we then have

$$g(x) = B_0 \sin x + 2 \sum B_k \cos kx \sin x,$$

with $B_k \downarrow 0$ ultimately. We apply Theorem 4.2, with index $\gamma - 1$, to $g(x)/\sin x = G(x)$, and infer that $\sum n^{\gamma-2} B_n$ converges. Since B_n is ultimately decreasing, $n^{\gamma-1} B_n \to 0$. Then by Lemma 2.6, with $y_k = B_k$ and $x_k = k^{\gamma-1}$, it follows that $\sum k^{\gamma-1} b_k$ converges.

Proof of Theorem 4.9, $\int \to \sum$. For $1 \leq \gamma \leq 2$ this follows in a similar way from Theorem 4.1; for $2 < \gamma < 3$, from Theorem 4.8.

We now prove Theorem 4.4; somewhat more generally, we show that the convergence of $\sum n^{\gamma-1} \lambda_n$ is necessary for the convergence of

(4.15) $$\int_{0+}^{\pi} x^{-\gamma} \varphi(x)\, dx$$

and sufficient for the convergence of

(4.16) $$\int_{a+}^{\pi} (x-a)^{-\gamma} \varphi(x)\, dx.$$

---

[1] O'SHEA [1], SHAH [1], BOAS [5].

Consider sine series. If (4.15) holds and $\varphi = g$, there is a number M such that

(4.17)
$$\left| \int_y^x t^{-\gamma} g(t) \, dt \right| \le M, \qquad 0 < y < x \le \pi.$$

It is easily verified that

$$y^{-\gamma} \int_0^y g(t) \, dt = \mathrm{o}(1), \qquad y \to 0.$$

Hence there is a number M' such that

(4.18)
$$\left| y^{-\gamma} \int_0^y g(t) \, dt \right| \le M'.$$

Since $\sum b_n \sin nx$ is a Fourier series, we may express the left-hand sides of (4.17) and (4.18) by integrating term by term; we get

(4.19)
$$\left| y^{-\gamma} \sum b_n \left\{ \int_0^y \sin nt \, dt + y^\gamma \int_y^x t^{-\gamma} \sin nt \, dt \right\} \right| \le M + M'.$$

The expression in braces in (4.19) is positive, since the second integral is (by the second mean-value theorem)

$$y^{-\gamma} \int_y^v \sin nt \, dt, \qquad y < v < x,$$

and therefore the whole expression is

$$\int_0^v \sin nt \, dt \ge 0.$$

Consequently for each N

$$y^{-\gamma} \sum_{n=1}^N b_n \left\{ \int_0^y \sin nt \, dt + y^\gamma \int_y^x t^{-\gamma} \sin nt \, dt \right\} \le M + M'.$$

Letting $y \to 0$, we have

$$\sum_{n=1}^N b_n \int_0^x t^{-\gamma} \sin nt \, dt \le M + M',$$

i.e.

$$\sum_{n=1}^N b_n n^{\gamma-1} \int_0^{nx} u^{-\gamma} \sin u \, du \le M + M'.$$

Take $x = 1$; the limit of the integral is $\Gamma(1-\gamma) \cos \frac{1}{2}\pi\gamma > 0$, and hence, since $b_n \ge 0$, $\sum n^{\gamma-1} b_n$ converges.

Suppose now that $\sum n^{\gamma-1} b_n$ converges. For each a,

$$\int_{a+x}^{a+y} (t-a)^{-\gamma} \sin nt \, dt \qquad (0 < x < y < \pi - a)$$

is easily seen to be uniformly $O(n^{\gamma-1})$. Consequently

(4.20)
$$\sum b_n \int_{a+x}^{a+y} (t-a)^{-\gamma} \sin nt \, dt$$

converges uniformly in x and y. Therefore (4.20) approaches a limit as $x \to 0+$. But (4.20) is, since $\sum b_n \sin nx$ is a Fourier series, equal to

$$\int_{a+x}^{a+y} (t-a)^{-\gamma} g(t)\, dt.$$

Hence

$$\int_{a+}^{a+y} (t-a)^{-\gamma} g(t)\, dt$$

converges.

The proof for cosine series is similar, except that we have to integrate one more time at the beginning, so that the analogue of (4.19) is

$$y^{-\gamma-1} \sum n^{-1} a_n \left\{ \int_0^y \sin nt\ dt + y^{\gamma+1} \int_y^x t^{-\gamma-1} \sin nt\ dt \right\} = O(1).$$

Some of the theorems of this section have been extended to larger values of γ, with differences of higher order replacing the functions. We quote the following theorems[1]).

Theorem 4.21. *Sines,* $\gamma \geq 2$. *Let* $b_n \geq 0$ *ultimately and let* b_n *be the sine coefficients of* g. *For an integer* $j \geq 1$ *let* $\sum n^{2j-1} b_n$ *converge. Then for* $2j < \gamma < 2(j+1)$ *the series* $\sum n^{\gamma-1} b_n$ *converges if and only if*

$$(4.22) \qquad x^{-\gamma} \left\{ g(x) - x\, g'(0) - \frac{x^3}{3!}\, g'''(0) - \cdots - \frac{x^{2j-1}}{(2j-1)!}\, g^{(2j-1)}(0) \right\} \in L.$$

For $\gamma = 2j$, $\sum n^{\gamma-1} (\log n)\, b_n$ *converges if and only if (4.22) holds.*

Theorem 4.23. *Cosines,* $\gamma \geq 3$. *Let* $a_n \geq 0$ *ultimately and let* a_n *be the cosine coefficients of* f. *For an integer* $j \geq 0$, *let* $\sum n^{2j} a_n$ *converge. Then for* $2j+1 < \gamma < 2(j+1)+1$ *the series* $\sum n^{\gamma-1} a_n$ *converges if and only if*

$$(4.24) \qquad x^{-\gamma} \left\{ f(x) - f(0) - \frac{x^2}{2!}\, f''(0) - \cdots - \frac{x^{2j}}{(2j)!}\, f^{(2j)}(0) \right\} \in L.$$

For $\gamma = 2j+1$, $\sum n^{\gamma-1} (\log n)\, a_n$ *converges if and only if (4.24) holds.*

Question 4.25. What conditions are necessary and sufficient for $\sum n^{\gamma-1} b_n$ or $\sum n^{\gamma-1} a_n$ to converge when $\gamma = 2j$ or $2j+1$?

Question 4.26. Extend Theorems 4.21 and 4.23 along the lines of Theorems 4.8 and 4.9, to involve integrability at points other than 0.

---

[1]) González-Fernández [1].

§ 5. The exceptional integral values of the index

Whereas the theorems of §§ 3 and 4 involve the function and the coefficients symmetrically when the index γ is not an integer, and sometimes when it is, there are a number of asymmetric theorems, and some missing ones, for integral values of γ, usually at values differing by 2. In this section we shall summarize the known facts and point out some connections with other aspects of the theory of trigonometric series, mentioning also a number of results that do not involve absolute convergence.

$\gamma = -1$. Theorem 3.1 shows that this value is exceptional for sine series. We know, for a positive g, a necessary and sufficient condition for $\sum n^{-2}|b_n|$ to converge, but not a condition for $x\, g(x) \in L$ (Theorem 3.1). There is nothing peculiar about cosine series for $\gamma = -1$.

$\gamma = 0$. For sine series, $\gamma = 0$ is not exceptional from the point of view of § 3 or § 4: we know that $\sum n^{-1}|b_n|$ converges if and only if $g \in L$ provided either that g decreases (Theorem 3.4) or b_n decreases (Theorem 4.1); in either case $b_n \geq 0$ and $g(x) \geq 0$ at least in a right-hand neighborhood of 0, so that we are dealing with a case of Parseval's theorem in the usual sense. The two theorems in this case considerably antedate most of the theory discussed here[1]); and both are special cases of general theorems of EDMONDS (see also § 8).

There are other theorems connecting the conditions $g \in L$ and $\sum n^{-1}|b_n| < \infty$.

Theorem 5.1. *If $b_n \to 0$ and*

$$(5.2) \qquad\qquad \sum |\varDelta\, b_n| \log n < \infty$$

then[2]) $g \in L$, $|b_n| \log n \to 0$, *and*

$$(5.3) \qquad\qquad \sum n^{-1}|b_n| < \infty\, ;$$

conversely if (5.3) *holds and* $|b_n| \log n \to 0$ *then* (5.2) *holds and* $g \in L$.

To show that (5.2) implies $g \in L$, we first observe that since

$$b_n = \sum_{k=n}^{\infty} \varDelta\, b_k$$

we have

$$\sum n^{-1}|b_n| \leq \sum n^{-1}\left| \sum_{k=n}^{\infty} \varDelta\, b_k \right| \leq \sum n^{-1} \sum_{k=n}^{\infty} |\varDelta\, b_k| = \sum |\varDelta\, b_k| \sum_{n=1}^{k} n^{-1}.$$

---

[1]) YOUNG [1], ZYGMUND [1].
[2]) SZIDON [1].

Then by partial summation it follows that $|b_n| \log n$ approaches a limit, which can only be 0. (Conversely (5.3) and $|b_n| \log n \to 0$ imply (5.2).) Now, again by partial summation,

$$\int \left| \sum_{k=m}^{n} b_k \sin kx \right| \le \sum_{k=m}^{n-1} |\varDelta b_k| \int |\tilde{D}_k(x)| \, dx + |b_n| \int |\tilde{D}_n(x)| \, dx + |b_{m-1}| \int |\tilde{D}_{m-1}(x)| \, dx,$$

where $\tilde{D}_n$ is the conjugate Dirichlet kernel. The integrals on the right are the Lebesgue constants for conjugate series, and the one with index k is of order $\log k$. Hence

$$\int \left| \sum_{k=m}^{n} b_k \sin kx \right| dx \le C \left\{ \sum_{k=m}^{n-1} |\varDelta b_k| \log k + |b_n| \log n + |b_{m-1}| \log (m-1) \right\}.$$

Since the right-hand side approaches 0 as m and $n \to \infty$, $\sum b_k \sin kx$ converges in L and hence is a Fourier series.

Theorem 5.4[1]). *If $b_n \to 0$ and $\{b_n\}$ is convex, or more generally*

$$(5.5) \qquad\qquad\qquad \sum n |\varDelta^2 b_n| < \infty,$$

then $g \in L$ if and only if (5.3) holds.

In Theorems 5.1 and 5.4, $\sum b_n \sin nx$ converges in virtue of (5.2) or (5.5).

The condition (5.2) is stronger than $\sum |\varDelta b_n| < \infty$, i.e. stronger than bounded variation, which is a natural generalization of $b_n \downarrow 0$. On the other hand, convexity is stronger than monotonicity, but its generalization (5.5) does not imply (5.2) (nor is it implied by (5.2)). TELJAKOVSKIĬ shows by examples that a $g(x)$ satisfying the hypotheses of Theorem 5.4 cannot necessarily be written as the difference of two integrable functions with decreasing sine coefficients, and that the convergence of (5.3) cannot be replaced by the convergence of $\sum n^{-1} b_n$.

We outline the proof of Theorem 5.4, since the method is interesting in itself. It is enough to show that for each integer s

$$\left| \int_{1/(s+1)}^{\pi} \left| \sum b_k \sin kx \right| dx - \sum_{k=1}^{s} k^{-1} |b_k| \right| \le C \sum k |\varDelta^2 b_{k-1}|.$$

By partial summation we can show that

$$- \sum_{k=1}^{\infty} b_k \sin kx = \sum_{k=1}^{\infty} \varDelta^2 b_{k-1} \frac{\sin kx}{4 \sin^2 \frac{1}{2}x}.$$

If we replace $\sin^2 \frac{1}{2}x$ by $x^2/4$, and $\sin kx$ by kx for $k \le 1/x$ and by 1 for $k \ge 1/x$, we commit an error that is easily seen to be bounded by the series in (5.5). It is consequently enough to show that on $1/(n+1) \le x \le 1/n$ $(n = 1, 2, ...)$ the quantity

$$\left| x^{-1} \sum_{k=1}^{n} k \, \varDelta^2 b_{k-1} + x^{-2} \sum_{k=n+1}^{\infty} \varDelta^2 b_{k-1} \right|$$

---

[1]) TELJAKOVSKIĬ [1].

differs from $|b_n|/x$ by something whose integral over $(1/(n+1), 1/n)$ is the nth term of a convergent series. But

$$\sum_{k=1}^{n} k \, \Delta^2 \, b_{k-1} = -b_n - n \, \Delta^2 \, b_n, \qquad \sum_{k=n+1}^{\infty} \Delta^2 \, b_{k-1} = \Delta \, b_n,$$

and so the difference in question is at most $|\Delta \, b_n|(x^{-2} - n \, x^{-1})$, whose integral is at most

$$|\Delta \, b_n| \le \sum_{k=n}^{\infty} |\Delta^2 \, b_k|,$$

which is indeed the nth term of a convergent series.

We next ask for the effect of weakening the hypothesis that g decreases by requiring only that g is positive. This question has already been answered by Theorem 3.8: if $x \, g(x) \in L$, $g(x) \ge 0$, and g is bounded outside a neighborhood of 0, then $g(x) \in L$ if and only if $\sum n^{-1} \, b_n$ converges.

We can also ask what happens when $b_n \downarrow 0$ is weakened to $b_n \ge 0$ and we suppose that b_n are the generalized sine coefficients of g.

Question 5.6. If $b_n \ge 0$ are the generalized sine coefficients of g, is it true that

(5.7) $$\int_{0+}^{\pi} g(x) \, dx \quad \text{exists}$$

if and only if

(5.8) $$\sum n^{-1} \, b_n \quad \text{converges?}$$

This is suggested by Theorem 4.4, but that theorem requires b_n to be Fourier sine coefficients in the first place.

Under other hypotheses, the conditions (5.7) and (5.8) are known to be equivalent.

Theorem 5.9[1]). *If $\sum b_n \sin nx$ converges boundedly on $(\delta, \pi)(\delta > 0)$ and*

$$\sum_{n}^{2n} |\Delta \, b_k| = O(1)$$

then (5.7) and (5.8) are equivalent.

Theorem 5.10. *If $\sum |\Delta \, b_n| < \infty$ then (5.7) and (5.8) are equivalent.*

If $\sum |\Delta \, b_n| < \infty$ then $\sum b_n \sin nx$ is a generalized sine series, since $\{b_n\}$ is the difference of two decreasing sequences. Consequently Theorem 5.10 is a special case of the following much more general result.

Theorem 5.11[2]). *If $x \, g(x) \in L$ and b_n are generalized sine coefficients of g then (5.7) holds if and only if $\sum n^{-1} \, b_n$ is summable $(C, -1+\delta)$ for all $\delta > 0$.*

We turn now to cosine theorems when $\gamma = 0$. Here our information, particularly about absolute convergence and absolute integrability, is very incomplete.

---

[1]) S. Izumi [1], [2].
[2]) This follows from 5.33 below by partial summation.

Question 5.12. If $a_n \downarrow 0$, so that $\sum a_n \cos nx$ converges to $f(x)$, what condition is necessary and sufficient for $f \in L$?

The following facts are known.

Theorem 5.13[1]). *If* $a_n \downarrow 0$, *or more generally if* $a_n \to 0$ *and* $\sum |\varDelta a_n| < \infty$, *then*

$$\int_{0+}^{\pi} f(x)\,dx \quad exists;$$

and $f \in L$ *if* $\sum |\varDelta a_n| \log n < \infty$, *or if* $a_n \downarrow 0$ *and* $\sum n^{-1} a_n < \infty$ *(the converse is false).*

Theorem 5.14[2]). *If* $a_n \downarrow 0$, $f \in L$ *if and only if* a_n *are Fourier-Stieltjes coefficients of a function of bounded variation.*

On the other hand, we know necessary and sufficient conditions for $\sum n^{-1} a_n$ to converge when[3]) $a_n \downarrow 0$ or when $f(x)$ decreases.

Theorem 5.15. *If* $a_n \downarrow 0$ *then* $\sum n^{-1} a_n$ *converges if and only if*

$$\int_{0+}^{\pi} f(x) \log(1/x)\,dx$$

converges.

This is a special case of Theorem 8.5 (below), since $\sum n^{-1} \cos nx = -\log(2 \sin \tfrac{1}{2}x) \sim \log(1/x)$ as $x \to 0+$.

Theorem 5.16. *If* f *decreases,* $f \in L$ *and* f *is bounded below, then* $\sum n^{-1} a_n$ *converges if and only if* $f(x) \log(1/x) \in L$.

This is the case $\gamma = 0$ of Theorem 3.5.

If a_n are Fourier coefficients without any further hypotheses, we also have a necessary and sufficient condition for $\sum n^{-1} a_n$ to converge.

Theorem 5.17[4]). *If* $f \in L$ *and* a_n *are its cosine coefficients, then* $\sum n^{-1} a_n$ *converges if and only if*

$$\textbf{(5.18)} \qquad \int_{-\delta}^{\delta} x^{-1}\,dx \int_{-x}^{x} f(t)\,dt \quad converges.$$

If $f(x) \geq 0$ then (5.18) is equivalent to $f(x) \log(1/x) \in L$, so Theorem 5.17 implies Theorem 5.16.

M. and S. Izumi [1] have shown that if

$$\log(1/x) \int_{0}^{x} |f(t)|\,dt \to 0$$

then $\sum n^{-1} a_n$ converges if and only if

$$\int_{0+}^{\pi} f(x) \log(1/x)\,dx \quad converges.$$

---

[1]) Szidon [1]; Zygmund [2], Chapter 5.

[2]) Goes [1].

[3]) See also Stečkin [1].

[4]) Hardy and Littlewood. See Hardy and Rogosinski [2], p. 96.

Cf. also ZYGMUND [2], vol. 1, p. 228, no. 8, and p. 286 (8.7) for sufficient conditions for the convergence of $\sum n^{-1} a_n$ and $\sum n^{-1}|a_n|$.

As a parallel to Theorem 5.16, we have the following result (BOAS [4]) about the generalized cosine coefficients of a positive function.

Theorem 5.19. *If* $f(x) \geq 0$, $x^2 f(x) \in L$ *and* a_n *are the generalized cosine coefficients of* f *then* $f \in L$ *if and only if*

$$n^{-1} \sum_{k=0}^{n} a_k$$

approaches a limit $\big(i.\ e.,\ a_k \to L(C, 1)\big)$.

We might also ask if the situation is improved by imposing a stronger condition on $\{a_n\}$. The natural condition is convexity, but then the problem disappears since if $a_n \to 0$ either $\Delta^2 a_n \geq 0$ or $\sum n|\Delta^2 a_n| < \infty$ implies that $\sum a_n \cos nx$ is a Fourier series (see ZYGMUND [2], vol. 1, p. 183). There are similar criteria involving differences of fractional order (CESARI. For references and additional results see BOAS [3]).

$\gamma = 1$. In a general way, we should expect a sine theorem for $\gamma = 1$ to correspond to a cosine theorem with $\gamma = 0$ and more stringent conditions of positivity, monotonicity, and so on; and similarly with "sine" and "cosine" interchanged. Actually not all theorems with one index correspond to significant theorems with the other, and we shall not in most cases attempt to trace out what equivalences there are. Since the natural series condition with $\gamma = 1$ is $\sum |a_n| < \infty$ or $\sum |b_n| < \infty$, we are in contact here with the whole theory of absolutely convergent Fourier series. We cannot expect too much in the way of absolute convergence of a Fourier series when we have a hypothesis of integrability, since absolute convergence is not a local property; we can succeed only if the coefficients are restricted in some way that forces the function to be very well behaved except in the neighborhood of a single point.

We consider sine series first.

Theorem 5.20. *If* $b_n \downarrow 0$ *then* $x^{-1} g(x) \in L$ *if and only if* $\sum b_n$ *converges.*

It is interesting to note that the additional condition

$$(\log n) \sum_{k=n}^{\infty} b_k \to 0$$

is necessary and sufficient for the Fourier series of $x^{-1} g(x)$ to converge in the L metric[1]).

Theorem 5.20 is a special case of Theorem 4.1. No dual theorem seems to be known.

Question 5.21. If g decreases on some $(0, \delta)$ and is bounded below, or if g is positive, and if $x g(x) \in L$ and b_n are the generalized sine coefficients of g, what are necessary and sufficient conditions for $x^{-1} g(x) \in L$ or for $\sum b_n$ to converge or for $\sum |b_n|$ to converge?

---

[1]) Partial summation applied to ZYGMUND [2], p. 184 (1.12).

Note that $x^{-1} g(x) \in L$ implies the convergence of $\sum b_n$ (not of $\sum |b_n|$) with no positivity hypothesis on g, by Dini's test for the convergence of the conjugate series (HEYWOOD [4]).

The next theorem is not symmetric, but has no positivity hypothesis at all (BOAS [1]).

Theorem 5.22. *If $\sum |b_n| < \infty$ then*

$$\int\limits_{0+}^{\pi} x^{-1} g(x)\, dx \quad \text{exists.}$$

Question 5.23. To what extent, if any, can Theorem 5.20 be extended to Fourier series with positive coefficients?

A partial answer is furnished by the next theorem[1]).

Theorem 5.24. *If $n\, b_n > -C$ (hence in particular if $b_n \geq 0$ ultimately), $\sum b_n = 0$ if and only if*

$$\int\limits_{0+}^{\to \infty} t^{-1} g(t)\, dt = 0.$$

For a counterexample to Theorem 5.20 when decreasing coefficients are replaced by positive coefficients see Theorem 5.27.

A simpler theorem of the same character as Theorem 5.24 still gives a necessary and sufficient condition for $\sum b_n$ to converge.

Theorem 5.25. *If $g \in L$ and its sine coefficients b_n are ultimately positive, $\sum b_n$ converges if and only if*

$$(5.26) \qquad \left| \int\limits_{0+}^{x} t^{-1} g(t)\, dt \right| \leq C, \quad 0 < x < \infty.$$

The $\sum \to \int$ part is elementary: if $\sum b_n$ converges and $\varepsilon > 0$,

$$\int\limits_{\varepsilon}^{x} t^{-1} g(t)\, dt = \sum_{n=1}^{\infty} b_n \int\limits_{\varepsilon}^{x} t^{-1} \sin nt\, dt = \sum_{n=1}^{\infty} b_n \int\limits_{n\varepsilon}^{nx} u^{-1} \sin u\, du.$$

Since the last integral is bounded, uniformly in ε and x, the series on the right converges uniformly. Let $\varepsilon \to 0$.

The $\int \to \sum$ part can be proved along the lines of Theorem 4.4, but a short proof is available from the summability theory of conjugate Fourier series. In fact, (5.26) is a sufficient condition for the Abel boundedness at 0 of the conjugate series of $\sum b_n \sin nx$. Hence $\sum b_n$ has bounded partial sums and so converges (because $b_n \geq 0$ ultimately).

The question of when $x^{-1} g(x) \in L$ can be answered completely when $\sum b_n \sin nx$ is lacunary (M. WEISS [1]).

---

[1]) HARDY and LITTLEWOOD [1].

Theorem 5.27. *If $g(x)$ has the Fourier series $\sum c_k \sin n_k x$, with $\sum c_k^2 < \infty$ and $n_{k+1}/n_k > q > 1$, then $x^{-1} g(x) \in L$ if and only if*

(5.28)
$$\sum_{m=1}^{\infty} \left(\log \frac{n_{m+1}}{n_m} \right) \left(\sum_{k=m+1}^{\infty} c_k^2 \right)^{\frac{1}{2}} < \infty.$$

This verifies the example $\sum n^{-3/2} \sin 2^n x$ given in connection with Theorem 5.20.

The proof of Theorem 5.27 is rather lengthy. Instead of reproducing it we shall establish the much simpler special case when $n_k = 2^k$; this illustrates the main ideas. It is more representative than might appear at first sight since we can always assume $n_{k+1}/n_k < r$ by adding terms with zero coefficients, and then (5.28) takes the form

(5.29)
$$\sum_{n=1}^{\infty} \left(\sum_{k=n+1}^{\infty} c_k^2 \right)^{\frac{1}{2}} < \infty.$$

I write $t(k)$ for 2^{-k} whenever this will simplify the typography.

$\sum \to \int$. This part can be given an elementary proof[1]. With an arbitrary N we have

$$\int_{\pi t(N+1)}^{\pi/2} |g(x)| \, x^{-1} \, dx = \sum_{n=1}^{N} \int_{\pi t(n+1)}^{\pi t(n)} \left| \left(\sum_{k=1}^{n} + \sum_{k=n+1}^{\infty} \right) c_k \sin 2^k x \right| x^{-1} \, dx \le P_N + Q_N.$$

Then

$$P_N \le \sum_{n=1}^{N} \sum_{k=1}^{n} \int_{\pi t(n+1)}^{\pi t(n)} |c_k| \, 2^k \, x \, x^{-1} \, dx$$

$$\le \pi \sum_{n=1}^{N} 2^{-n-1} \sum_{k=1}^{n} |c_k| \, 2^k \le \pi \sum_{k=1}^{N} 2^k |c_k| \sum_{n=k}^{\infty} 2^{-n-1} = \pi \sum_{k=1}^{N} |c_k|,$$

so that P_N is bounded if $\sum |c_k| < \infty$, and so in particular if (5.29) holds.

We also have

$$Q_N \le \sum_{n=1}^{N} \int_{\pi}^{2\pi} \left| \sum_{k=n+1}^{\infty} c_k \sin 2^{k-n} x \right| x^{-1} \, dx$$

$$\le \sum_{n=1}^{N} \left\{ \int_{\pi}^{2\pi} \left| \sum_{k=n+1}^{\infty} c_k \sin 2^{k-n} x \right|^2 dx \right\}^{\frac{1}{2}} \left\{ \int_{\pi}^{2\pi} x^{-2} \, dx \right\}^{\frac{1}{2}} \le C \sum_{n=1}^{N} \left\{ \sum_{k=n+1}^{\infty} |c_k|^2 \right\}^{\frac{1}{2}}.$$

Hence Q_N is also bounded if (5.29) holds, and so (5.29) implies $\int |g(x)| x^{-1} \, dx < \infty$.

$\int \to \sum$. If we can show that Q_N is bounded then (5.29) follows, since

$$Q_N = \sum_{n=1}^{N} \int_{\pi t(n+1)}^{\pi t(n)} \left| \sum_{k=n+1}^{\infty} c_k \sin 2^k x \right| x^{-1} \, dx = \sum_{n=1}^{N} \int_{\pi}^{2\pi} \left| \sum_{k=n+1}^{\infty} c_k \sin 2^{k-n} x \right| x^{-1} \, dx$$

$$\ge C \sum_{n=1}^{N} \int_{\pi}^{2\pi} \left| \sum_{k=n+1}^{\infty} c_k \sin 2^{k-n} x \right| = C \sum_{n=1}^{N} \int_{0}^{2\pi} \left| \sum_{k=n+1}^{\infty} c_k \sin 2^{k-n} x \right| dx.$$

---

[1] I am indebted to S. Izumi for this observation.

By (8.20) of Zygmund [2], vol. 1, p. 215, the last expression exceeds

$$\text{(5.30)} \qquad\qquad C \sum_{n=1}^{N} \left\{ \sum_{k=n+1}^{\infty} c_k^2 \right\}^{\frac{1}{2}},$$

as required. (The case of general n_k requires a more sophisticated lemma of the same kind.)

Now we have

$$\int_{\pi t(N+1)}^{\pi/2} |g(x)|\, x^{-1}\, dx \geq Q_N - P_N.$$

If $\sum |c_k| < \infty$ then P_N is bounded, as we showed in the first part of the proof, and so Q_N is bounded if $\int |g(x)|\, x^{-1}\, dx$ is finite. If $\sum |c_k| = \infty$ we proceed to show that $P_N = o(Q_N)$ as $N \to \infty$, so if N is large enough we again have Q_N bounded. Since Q_N exceeds (5.30) and

$$P_N \leq \pi \sum_{k=1}^{N} |c_k|,$$

we need only show that

$$\text{(5.31)} \qquad\qquad \sum_{k=1}^{N} |c_k| = o\left\{ \sum_{n=1}^{N} \left(\sum_{k=n+1}^{\infty} c_k^2 \right)^{\frac{1}{2}} \right\}, \qquad N \to \infty.$$

Since rearranging $\{|c_k|\}$ in decreasing order increases the left-hand side and decreases the right-hand side of (5.31), it is enough to prove (5.31) with $|c_k|$ decreasing. Take a large (fixed) M. Then

$$\sum_{n=1}^{N} \left\{ \sum_{k=n+1}^{\infty} c_k^2 \right\}^{\frac{1}{2}} \geq \sum_{n=1}^{N} \left\{ \sum_{k=n+1}^{n+M} c_k^2 \right\}^{\frac{1}{2}}$$

$$\geq \sum_{k=1}^{N} M^{\frac{1}{2}} |c_{k+M}| \geq M^{\frac{1}{2}} \left(\sum_{k=1}^{N} |c_k| - \sum_{k=1}^{M-1} |c_k| \right).$$

Since $\sum |c_k|$ diverges and M can be arbitrarily large, (5.31) follows.

For cosine series we should not expect $\int x^{-1} |f(x)|\, dx$ and $\sum |a_n|$ to be connected with each other. We already know a condition for integrability (Theorem 4.9):

Theorem 5.32. *If $a_n \geq 0$ ultimately, and a_n are the cosine coefficients of f, then $\int x^{-1} |f(x) - f(0)| < \infty$ if and only if $\sum a_k \log k$ converges.*

Let us notice that the convergence of $\sum a_k \log k$ is, when $a_n \geq 0$ ultimately, equivalent to the convergence of $\sum n^{-1} s_n$, where $s_n = a_1 + a_2 + \cdots + a_n$. This suggests considering the series $\sum n^{-1} s_n$ in other cases; we quote the following theorem on conditional convergence[1].

Theorem 5.33. *If $f \in L$, a_n are the cosine coefficients of f, and*

$$\int_0^x |f(t)|\, dt = O(x)$$

---

[1] Bosanquet and Offord [1].

$(i.e., |f(x)| = O(1)(C, 1))$, then $\sum n^{-1} s_n$ is summable $(C, -1+\delta)$ (in particular, converges) if and only if

$$\int_{0+}^{x} t^{-1} f(t) \, dt$$

converges.

Theorems about $\sum |a_n|$ involve generalizations of boundedness or approach to a limit for f, rather than the integrability of f (cf. § 7). This fact indicates how extremely restrictive it is to require that a cosine series has positive coefficients: a mild additional restriction on the behavior of $f(x)$ at 0 makes the series converge absolutely.

Theorem 5.34[1]). If $f \in L$, if a_n are the cosine coefficients of f, and $a_n \geq 0$, then

(5.35) $$\int_{0}^{x} dt \int_{0}^{t} f(u) \, du \leq O(x^2)$$

is equivalent to $\sum a_n < \infty$.

Condition (5.35) says that $f(x)$ is bounded above in the $(C, 2)$ sense.
Since the left-hand side of (5.35) is

$$\sum n^{-2} a_n \sin^2 \tfrac{1}{2} nx \geq 0,$$

we could equally well take the hypothesis in the form $f(x) = O(1)(C, 2)$; and we could deduce the conclusion from general facts about Riemann summability.

The next theorem[2]) is a deeper result of the same kind.

Theorem 5.36. If $f \in L$, if a_n are the cosine coefficients of f, and if $n a_n > -C$, then $\sum a_n = 0$ if and only if

$$\int_{0}^{x} f(t) \, dt = o(x)$$

$(i.e., f(x) \to 0(C, 1))$.

---

[1]) Implicit in SzÁSZ [1]. It includes Paley's theorem that if a cosine series with positive coefficients is the Fourier series of a bounded function then it converges absolutely (HARDY and ROGOSINSKI [2], p. 72).

[2]) HARDY and LITTLEWOOD [1].

§ 6. L^p problems, $1 < p < \infty$

We now consider conditions for $x^{-\gamma} \varphi(x)$ to belong to L^p, $1 < p < \infty$, where φ stands for f or g and λ_n are its associated Fourier coefficients.

For $\gamma = 0$ there is a classical result of Hardy and Littlewood[1]).

Theorem 6.1. *If* $\lambda_n \downarrow 0$ *and* $1 < p < \infty$ *then* $\varphi \in L^p$ *if and only if* $\sum n^{p-2} \lambda_n^p < \infty$.

Here the sine and cosine theorems are equivalent since f and g are conjugate functions and so either one is in L^p if the other is.

There is a dual theorem (also Hardy and Littlewood's) with decreasing function instead of decreasing coefficients[2]).

Theorem 6.2. *If* $\varphi(x) \geq 0$ *and* φ *decreases,* $1 < p < \infty$, *and* λ_n *are the Fourier sine or cosine coefficients of* φ, *then* $\sum |\lambda_n|^p < \infty$ *if and only if* $x^{p-2} \varphi(x)^p \in L$.

These theorems can be extended to other values of γ.

Theorem 6.3[3]). *If* $\lambda_n \downarrow 0$ *then* $x^{-\gamma} \varphi(x) \in L^p$, $-1/p' < \gamma < 1/p$, *if and only if* $\sum n^{p\gamma + p - 2} \lambda_n^p$ *converges.* ·

When $\gamma = (2/p) - 1 = (1/p) - (1/p')$, Theorem 6.3 has the same form as Theorem 6.2 except that $\{\lambda_n\}$ instead of φ decreases.

Note that the range of γ in Theorem 6.3 is still, as in § 4, a unit interval, but shifted downward as p increases.

Question 6.4. What happens when $\gamma = -1/p'$ or $\gamma = 1/p$?

Theorem 6.5. *If* $\varphi(x) \geq 0$ *and* φ *decreases,* $1 < p < \infty$, $-1/p' < \gamma < 1/p$, *then* $\sum n^{-\gamma p} |\lambda_n|^p$ *converges if and only if* $x^{p\gamma + p - 2} \varphi(x)^p \in L$.

This can be proved by dualizing one of the proofs of Theorem 6.3. If we put $\gamma + 1 - 2/p = -\delta$, we can restate Theorem 6.5 by saying that $\sum n^{p\delta + p - 2} |\lambda_n|^p$ converges if and only if $x^{-\delta} \varphi(x) \in L^p$, where $-1/p' < \delta < 1/p$; this is precisely Theorem 6.3 with decreasing coefficients replaced by a decreasing function.

Theorems 6.3 and 6.5 suggest extensions to the Lorentz spaces[4]) $L(q, p)$. We say that $\varphi \in L(q, p)$ $(1 < p < \infty, 1 < q < \infty)$ if $t^{(1/q) - (1/p)} \varphi^* \in L^p$, where φ^* is φ

---

[1]) See ZYGMUND [2], vol. 2, p. 129.

[2]) ZYGMUND [2], vol. 2, p. 130.

[3]) CHEN [1], [2]. Other proofs are given by ASKEY and WAINGER [1] and BOAS [7].

[4]) See for example LORENTZ [2].

rearranged in decreasing order, i.e. φ^1) is the decreasing function equimeasurable with φ. Similarly $\{\lambda_n\} \in l(q, p)$ if $n^{(1/q)-(1/p)} \lambda_n^* \in l^p$.

Theorem 6.6. *If $\lambda_n \downarrow 0$ or if $\varphi(x) \geq 0$ and φ decreases, $1 < p < \infty$, $1 < q < \infty$, then $\varphi \in L(q, p)$ if and only if $\{\lambda_n\} \in l(q', p)$.*

This is stated for decreasing coefficients (also for $p = 1$ and ∞) by DIKAREV and MACAEV [1]. The cases $q = p$ and $q = p'$ reduce to theorems 6.1 and 6.2, and the proofs are not essentially different in the general case.

As we would expect from § 4, we can get along with positive λ_n when γ is larger; however, the condition on the coefficients has a different form (equivalent to that in Theorem 6.3 when the λ_n decrease).

Theorem 6.7[1]). *If λ_n are the Fourier sine or cosine coefficients of the continuous function φ, $1 < p < \infty$, $1/p < \gamma < (1/p)+1$, and $\lambda_n \geq 0$, then*

$$(6.8) \qquad |x - a|^{-\gamma} |\varphi(x) - \varphi(a)| \in L^p \text{ for every } a, 0 \leq a < \pi,$$

if and only if

$$(6.9) \qquad \sum n^{p\gamma - p - 2} \left(\sum_{k=1}^{n} k \lambda_k \right)^p < \infty,$$

or equivalently

$$(6.10) \qquad \sum n^{p\gamma - 2} \left(\sum_{k=n}^{\infty} \lambda_k \right)^p < \infty.$$

More precisely, (6.9) is necessary for (6.8) with $a = 0$, and sufficient for (6.8) with any a.

There is also a result that corresponds to Theorem 6.7 as Theorem 6.5 corresponds to Theorem 6.3. We state only the sine theorem[2]).

Theorem 6.11. *If $g(x) \geq 0$ and b_n are the generalized sine coefficients of g, if $1 < p < \infty$ and if $1/p < \gamma < 1+1/p$, then $\sum n^{-\gamma p} |b_n|^p$ converges if and only if*

$$x^{\gamma - 1 - 2/p} \int_0^x t g(t) \, dt \in L^p.$$

If we again put $-\delta = \gamma + 1 - 2/p$, we can state the result in the form: $\sum n^{p\delta + p - 2} |b_n|^p$ converges *if and only if*

$$x^{-\delta - 2} \int_0^x t g(t) \, dt \in L^p,$$

$-1/p'-1 < \delta < -1/p'$; this makes the theorem appear as an extension of Theorems 6.3 and 6.5 (in the second formulation) to still smaller values of the index.

We now return to smaller values of γ and ask about the possibility of weakening the condition $\lambda_n \downarrow 0$ in Theorem 6.3. (Cf. also § 8.)

---

[1]) BOAS [7].

[2]) See ASKEY and BOAS [1] for this and related results.

Question 6.12. What condition is necessary and sufficient for $x^{-\gamma} \varphi(x) \in L^p$, $-1/p' < \gamma < 1/p$, when $\lambda_n \geq 0$?

The answer to 6.12 is not the convergence of $\sum n^{p\gamma+p-2} \lambda_n^p$, at least for $p > 2/(1+\gamma)$. A counterexample (R. ASKEY) is $\varphi(x) = \sum k^{-\alpha} \cos 2^k x$ or $\sum k^{-\alpha} \sin 2^k x$, $\frac{1}{2} < \alpha \leq 1$. For (ZYGMUND [2], vol. 1, p. 215) φ belongs to every L^p with $p > 0$. By Hölder's inequality with exponent q satisfying $1 < q < 1/(\gamma p)$,

$$\int |x^{-\gamma} \varphi(x)|^p \, dx = \int x^{-\gamma p} |\varphi(x)|^p \, dx \leq \left(\int x^{-\gamma p q} \, dx \right)^{1/q} \left(\int |\varphi(x)|^{p q'} \, dx \right)^{1/q'} < \infty.$$

On the other hand $\sum n^{p\gamma-p+2} \lambda_n^p = \sum 2^{k(p\gamma+p-2)} k^{-\alpha p}$, which diverges if $p\gamma + p - 2 > 0$.

When $\lambda_n \geq 0$[1]), a necessary condition for $\varphi(x) x^{-\gamma} \in L^p$ is

$$\sum n^{p\gamma-2} \left(\sum_{k=0}^{n} \lambda_k \right)^p < \infty,$$

or equivalently

$$\sum n^{p\gamma+p-2} \left(\sum_{k=n}^{\infty} k^{-1} \lambda_k \right)^p < \infty;$$

and a sufficient condition is

$$\sum n^{p+p\gamma-2} \left(\sum_{k=n}^{\infty} |\lambda_k - \lambda_{k+2}| \right)^p < \infty.$$

ROBERTSON [2] has proved some L^p theorems for the case when $\{\lambda_n\}$ can be partitioned into k monotonic sequences.

We shall give proofs of Theorems 6.3 and 6.7. We need the following generalized form of Hardy's inequality (HARDY, LITTLEWOOD and PÓLYA [1], Theorem 330):

Lemma 6.13. If $p > 1$, $r > 1$, and ψ is integrable and positive, then

(6.14)
$$\int x^{-r} \left(\int_0^x \psi(t) \, dt \right)^p \leq K \int x^{-r} (x \psi(x))^p \, dx.$$

The best value of K is $\{p/(r-1)\}^p$, but we do not need this.

We shall use Lemma 6.13 when $r = p$ (the original form of Hardy's inequality); and when $r = p\gamma + p$, $\gamma > -1/p'$, in which case (6.14) says that

$$x^{-\gamma-1} \int_0^x f(t) \, dt \in L^p$$

if $f \in L^p$.

We also need a variant of Hardy's inequality for series (HARDY, LITTLEWOOD and PÓLYA [1], Theorem 346):

Lemma 6.15. If $m < 1$, $\lambda_k \geq 0$, $p > 1$,

$$\sum n^{-m} \left(\sum_{k=n}^{\infty} \lambda_k \right)^p \leq K \sum n^{-m} (n \lambda_n)^p.$$

---

[1]) ASKEY and WAINGER [1]; see also BOAS [7].

Lemma 6.16. *If $\gamma < 2$,*

$$t^{\gamma-2} \int_0^t u^{-\gamma} \sin u \, du$$

decreases on $(0, \pi)$.

Lemma 6.17. *If $\gamma < 3$,*

$$t^{\gamma-3} \int_0^t u^{-\gamma} (1 - \cos u) \, du$$

decreases on $(0, \pi)$.

The proofs of Lemmas 6.16 and 6.17 are straightforward applications of calculus.

Lemma 6.18. *If $\lambda_k \geq 0$, $p > 1$, $s > 0$, then*

(6.19)
$$\sum_{n=1}^{\infty} n^c \left(\sum_{k=n}^{\infty} \lambda_k \right)^p < \infty$$

implies

(6.20)
$$\sum_{n=1}^{\infty} n^{c-sp} \left(\sum_{k=1}^{n} k^s \lambda_k \right)^p < \infty$$

if $c < sp - 1$; if $c > -1$ then (6.19) implies (6.18).

Hence if $s > 0$, (6.19) and (6.20) are equivalent for $-1 < c < sp-1$.

First suppose (6.19) holds. Write

$$\Lambda_n = \sum_{k=n}^{\infty} \lambda_k;$$

then the left-hand side of (6.20) is

$$\sum_{n=1}^{\infty} n^{c-sp} \left\{ \sum_{k=1}^{n} k^s (\Lambda_k - \Lambda_{k+1}) \right\}^p.$$

By partial summation, the left-hand side of (6.20) does not exceed

(6.21)
$$s^p \sum_{n=1}^{\infty} n^{c-sp} \left(\sum_{k=1}^{n} k^{s-1} \Lambda_k \right)^p.$$

If $c - sp < -1$, Lemma 6.15 shows that (6.21) does not exceed

$$K \sum_{n=1}^{\infty} n^{c-sp} (n^s \Lambda_n)^p = K \sum_{n=1}^{\infty} n^c \Lambda_n^p < \infty.$$

Now suppose that (6.20) holds. We begin by showing that $\sum \lambda_k$ converges. Put

$$\Lambda_n = \sum_{k=1}^{n} k^s \lambda_k.$$

Then

(6.22)
$$\sum_{k=n}^{N} \lambda_k = \sum_{k=n}^{N} k^{-s} (\Lambda_k - \Lambda_{k-1}) \leq s \sum_{k=n}^{N-1} k^{-s-1} \Lambda_k + \Lambda_N N^{-s},$$

by partial summation. The sum on the right is

$$\sum_{k=n}^{N-1} \Lambda_k \, k^{(c/p)-s} \, k^{-1-(c/p)} \leq \left(\sum_{k=n}^{N-1} \Lambda_k^p \, k^{c-ps} \right)^{1/p} \left(\sum_{k=n}^{N-1} k^{-(p+c)/(p-1)} \right)^{1/p'}$$

by Hölder's inequality; the right-hand side has a bound independent of N because (6.20) holds and $(p+c)/(p-1) > 1$. Hence the first term on the right of (6.22) is bounded. The second term on the right of (6.22) approaches zero since

$$n^{-s} \Lambda_n \leq C \Lambda_n \sum_{k=n}^{2n} k^{-s-1} \leq C \sum_{k=n}^{2n} k^{-s-1} \Lambda_k \leq C \sum_{k=n}^{\infty} k^{-s-1} \Lambda_k,$$

and we have just shown that the last series converges. Hence $\sum \lambda_k$ converges, and

$$\sum_{k=n}^{\infty} \lambda_k \leq s \sum_{k=n}^{\infty} k^{-s-1} \Lambda_k.$$

Now by Lemma 6.15 again,

$$\sum_{n=1}^{\infty} n^c \left(\sum_{k=n}^{\infty} \lambda_k \right)^p \leq K \sum_{n=1}^{\infty} n^{c-sp} \Lambda_n^p < \infty.$$

This completes the proof of Lemma 6.18.

We now prove Theorem 6.7 for sine series. Let $g(x) \, x^{-\gamma} \in L^p$. By Theorems 4.4 and 4.8, $\sum n^{\gamma-1} \lambda_n$ converges. Now if $\gamma > 0$ and $\varepsilon > 0$,

$$\left| \int_{\varepsilon}^{x} t^{-\gamma} \sin nt \, dt \right| = n^{\gamma-1} \left| \int_{n\varepsilon}^{nx} u^{-\gamma} \sin u \, du \right| \leq C \, n^{\gamma-1},$$

with C independent of x and ε. Hence

$$\int_{\varepsilon}^{x} t^{-\gamma} g(t) \, dt = \sum \lambda_n \int_{\varepsilon}^{x} t^{-\gamma} \sin nt \, dt$$

converges uniformly in ε. Letting $\varepsilon \to 0$, we obtain

(6.23) $$\int_{0}^{x} t^{-\gamma} g(t) \, dt = \sum \lambda_n \int_{0}^{x} t^{-\gamma} \sin nt \, dt$$

if $\gamma > 0$; and each term on the right is positive.

Now by applying Lemma 6.13 to the left-hand side of (6.23) we get

$$\int_{0}^{\pi} dx \left(x^{-1} \int_{0}^{x} t^{-\gamma} |g(t)| \, dt \right)^p \leq K \int_{0}^{\pi} |t^{-\gamma} g(t)|^p \, dt < \infty.$$

Hence

$$\int_{0}^{\pi} \left(\sum \lambda_n x^{-1} \int_{0}^{x} t^{-\gamma} \sin nt \, dt \right)^p < \infty,$$

which we can write as

$$\int_{0}^{\pi} dx \left(\sum n \lambda_n x^{1-\gamma} (n x)^{\gamma-2} \int_{0}^{nx} u^{-\gamma} \sin u \, du \right)^p < \infty.$$

We decrease the left-hand side by replacing $\sum$ by $\sum_{1}^{1/x}$ and then replacing

$$(n\,x)^{\gamma-2}\int_0^{nx} u^{-\gamma}\sin u\;du$$

by its minimum (Lemma 6.16). Hence

$$\int_0^\pi \left(\sum_{n=1}^{1/x} n\,\lambda_n\,x^{1-\gamma}\right)^p dx < \infty,$$

or equivalently

$$\sum n^{p\gamma-p-2}\left(\sum_{k=1}^{n} k\,\lambda_k\right)^p < \infty.$$

For the converse, we prove somewhat more, namely that (6.9) and (6.10) for $|\lambda_n|$ imply (6.8) whether the λ_n are positive or not.

First suppose $a = 0$. With any $N > 0$,

$$\left\{\int_0^\pi \left|t^{-\gamma}\sum_{k=1}^{N}\lambda_k \sin kt\right|^p dt\right\}^{1/p} \le \left\{\int_0^\pi \left|t^{-\gamma}\sum_{k=1}^{1/t}\lambda_k \sin kt\right|^p dt\right\}^{1/p}$$

$$+\left\{\int_0^\pi \left|t^{-\gamma}\sum_{k=1/t}^{N}\lambda_k \sin kt\right|^p dt\right\}^{1/p} = S_1 + S_2.$$

Then

$$S_1 \le \int_0^\pi \left(t^{-\gamma+1}\sum_{k=1}^{1/t} k|\lambda_k|\right)^p dt \le \sum_{n=1}^\infty n^{p\gamma-p-2}\left(\sum_{k=1}^{n} k|\lambda_k|\right)^p,$$

and

$$S_2 \le \int_0^\pi \left(t^{-\gamma}\sum_{k=1/t}^\infty |\lambda_k|\right)^p dt \le \sum_{n=1}^\infty n^{\gamma p-2}\left(\sum_{k=n}^\infty |\lambda_k|\right)^p.$$

More generally,

$$g(x)-g(a) = 2\sum \lambda_n \cos \tfrac{1}{2} n(x+a) \sin \tfrac{1}{2} n(x-a),$$

and the same argument applies. It also applies to cosine series, since

$$f(x)-f(a) = -2\sum \lambda_n \sin \tfrac{1}{2} n(x+a) \sin \tfrac{1}{2} n(x-a).$$

The $\int \to \sum$ part of the cosine form of Theorem 6.7 can be proved by a slightly simpler argument. Assume that there is a number s such that $x^{-\gamma}[f(x)-s] \in L^p$. Since $\gamma > 1/p$, this implies $x^{-1}[f(x)-s] \in L$, so that $\sum \lambda_n \cos nx$ converges at 0 and $s = f(0)$. Then

$$f(0)-f(x) = \sum \lambda_n(1-\cos nx).$$

Since the series has positive terms it can be integrated term by term to get

$$\int_0^x t^{-\gamma}[f(0)-f(t)]\;dt = \sum \lambda_n \int_0^x t^{-\gamma}(1-\cos nt)\,dt.$$

By Lemma 6.13,

$$\int_0^\pi \left(\sum \lambda_n x^{-1} \int_0^x t^{-\gamma}(1-\cos nt)\, dt \right)^p dx \le \int_0^\pi dx \left(x^{-1} \int_0^x t^{-\gamma} |f(0) - f(t)|\, dt \right)^p$$

$$\le K \int_0^\pi t^{-\gamma} |f(0) - f(t)|^p \, dt.$$

Hence the left-hand side is finite; hence so is

$$\int_0^\pi \left\{ \sum_{n=0}^{1/x} \lambda_n x^{-1} n^{\gamma-1} \int_0^{nx} u^{-\gamma}(1-\cos u)\, du \right\}^p dx$$

$$= \int_0^\pi \left\{ \sum_{n=0}^{1/x} n^2 \lambda_n x^{2-\gamma} (n x)^{\gamma-3} \int_0^{nx} u^{-\gamma}(1-\cos u)\, du \right\}^p dx,$$

and hence so (by Lemma 6.17) is

$$\int_0^\pi \left\{ \sum_{n=0}^{1/x} n^2 \lambda_n x^{2-\gamma} \right\}^p dx.$$

This is equivalent to

$$\sum n^{p\gamma - 2p - 2} \left(\sum_{k=1}^n k^2 \lambda_k \right)^p < \infty,$$

and by Lemma 6.18 $(s = 2, c = p\gamma - 2)$ to

$$\sum n^{p\gamma - 2} \left(\sum_{k=n}^\infty \lambda_k \right)^p < \infty.$$

We now deduce the cosine form of Theorem 6.3 from the sine form of Theorem 6.7; the sine form of Theorem 6.3 is obtained similarly.

First let $-1/p' < \gamma < 1/p$, $f(x) x^{-\gamma} \in L^p$, and let the cosine coefficients of f be $a_n \downarrow 0$ ultimately. (Since $\gamma > -1/p'$, $x^{-\gamma} f(x) \in L^p$ implies $f \in L$.) Suppose for simplicity that $a_0 = 0$. We have

$$f_1(x) = \int_0^x f(t)\, dt = \sum n^{-1} a_n \sin nx.$$

Apply Theorem 6.7 to f_1 with $\gamma + 1$ in place of γ. Lemma 6.13 shows that $x^{-\gamma-1} f_1(x) \in L^p$ and consequently

$$\sum n^{p\gamma - 2} \left(\sum_{k=0}^n a_k \right)^p < \infty.$$

Since $a_k \downarrow$, we have successively

$$\sum n^{p\gamma - 2} \left(\sum_{k=n/2}^n a_k \right)^p < \infty,$$

$$\sum n^{p\gamma-2}\left(a_n \sum_{k=n/2}^{n} 1\right)^p < \infty,$$

$$\sum n^{p\gamma+p-2} a_n^p < \infty.$$

Next suppose that $a_n \downarrow 0$ and $f(x) = \sum a_n \cos nx$. Then

$$2 f(x) \sin x = \sum (a_n - a_{n+2}) \sin(n+1)x.$$

Hence by Theorem 6.7 with $\gamma+1$ in place of γ, $x^{-\gamma-1}\left(x\, f(x)\right) \in L^p$, i. e., $x^{-\gamma} f(x) \in L^p$, if

$$\sum n^{p\gamma+p-2}\left(\sum_{k=n}^{\infty} |a_k - a_{k+2}|\right)^p \quad \text{converges.}$$

But

$$\sum_{k=n}^{\infty} |a_k - a_{k+2}| = a_n + a_{n+1},$$

so $x^{-\gamma} f(x) \in L^p$ if $\sum n^{p\gamma+p-2} a_n^p$ converges.

§ 7. Asymptotic formulas and Lipschitz conditions

If we let $p \to \infty$ in the theorems of § 6 they go over formally into theorems about the boundedness of $x^{-\gamma} \varphi(x)(-1 < \gamma < 0)$ or of $|x-a|^{-\gamma}|\varphi(x)-\varphi(a)| \ (0 < \gamma < 1)$, and about the boundedness of $n^{\gamma+1} \lambda_n$ or of

$$n^{\gamma} \sum_{k=n}^{\infty} \lambda_k.$$

The theorems suggested in this way are true, some of them in more precise forms; most of them were originally discovered independently of the theory discussed in § 6.

When $0 < \gamma < 1$ we are dealing with functions φ in the class Lip γ. When $-1 < \gamma < 0$ it is convenient to write $\alpha = \gamma+1$, so that we are considering connections between $\varphi(x) = O(x^{\alpha-1})$, $0 < \alpha < 1$, and $\lambda_n = O(n^{-\alpha})$. In this case there are more precise asymptotic formulas.

Theorem 7.1[1]). *If* $b_n \downarrow 0$ *and* $0 < \alpha < 2$,

$$(7.2) \qquad\qquad g(x) \sim K x^{\alpha-1}, \quad x \to 0+;$$

if and only if

$$(7.3) \qquad\qquad b_n \sim (2/\pi) K \, \Gamma(\alpha) \, \sin \tfrac{1}{2}\pi\alpha \, n^{-\alpha}, \quad n \to \infty.$$

Theorem 7.4. *If* $a_n \downarrow 0$ *and* $0 < \alpha < 1$,

$$(7.5) \qquad\qquad f(x) \sim K x^{\alpha-1}, \quad x \to 0+,$$

if and only if

$$(7.6) \qquad\qquad a_n \sim (2/\pi) K \, \Gamma(\alpha) \, \cos \tfrac{1}{2}\pi\alpha \, n^{-\alpha}, \quad n \to \infty.$$

If $K = 0$, (7.2) is to be interpreted as $g(x) = o(x^{\alpha-1})$, and correspondingly for the other formulas. There are weaker forms in which (7.2) is replaced by $g(x) = O(x^{\alpha-1})$, and so on.

There are many generalizations and analogues of Theorems 7.1 and 7.4 when the powers are replaced by more general functions; see HARDY and ROGOSINSKI [3], where an extensive bibliography is given; see also ZYGMUND [2], Chapter 5; ALJANČIĆ, BOJANIĆ and TOMIĆ [2]. What is most interesting from our present point of view is to consider the possibility of extending the range of α and of weakening the hypothesis that the coefficients are monotonic.

---

[1]) For this and the next theorem see HARDY [2], [3], HEYWOOD [2].

Let us first note that there is a dual of Theorem 7.1 when the function decreases.

Theorem 7.7[1]). *If $g(x)\downarrow$, g is bounded below, b_n are the Fourier coefficients of g, and $0 < \alpha < 1$, then (7.2) holds.*

Question 7.8. Obtain an analogue of Theorem 7.4 when $f\downarrow$.

If $\alpha \geq 1$, (7.2) is incompatible with $g(x)\downarrow$. However, the general pattern of previous theorems suggests that the correct extension of Theorem 7.7 is to $-1 < \alpha < 0$.

Question 7.9. Extend Theorem 7.7 to the range $-1 < \alpha < 0$, with b_n generalized Fourier coefficients.

Previous theorems also suggest that, in the first place, the hypothesis that b_n is decreasing is not really essential in Theorem 7.2; second, that whereas Theorem 7.4 fails when $\alpha \geq 1$, there should be an extension to $1 < \alpha < 3$ involving $f(x) - f(0)$ instead of $f(x)$. It is natural, when $\alpha > 1$, to start with the assumption that the series are Fourier series, since (7.3) or (7.6), when $\alpha > 1$, makes the series converge absolutely.

Theorem 7.10[2]). *If $b_n \geq 0$, b_n are the sine coefficients of g, and $1 < \alpha < 2$, then*

(7.11)
$$g(x) \sim K\, x^{\alpha - 1}, \quad x \to 0+,$$

if and only if

(7.12)
$$B_n \sim (K/\pi)\, \Gamma(\alpha - 1)\, \sin \tfrac{1}{2}\pi\alpha\; n^{1-\alpha},$$

where

(7.13)
$$B_n = \sum_{j=0}^{\infty} b_{2j+n+1}.$$

If $b_n \downarrow 0$, the part of Theorem 7.1 with $1 < \gamma < 2$ can be deduced from Theorem 7.10 by an elementary Tauberian argument.

We could replace (7.13) by the slightly more transparent but slightly less general

$$B_n = \sum_{k=n}^{\infty} b_k.$$

Theorem 7.14. *If $a_n \geq 0$, a_n are the cosine coefficients of f, and $1 < \alpha < 3$, then*

(7.15)
$$f(x) - c \sim K\, x^{\alpha - 1}, \quad x \to 0+,$$

if and only if $\sum a_n$ converges, $c = \tfrac{1}{2}a_0 + \sum a_n$, and

---

[1]) HARDY [3].
[2]) For this and the next theorem see BOAS [6].

(7.16) $$A_n \sim (K/\pi)\, \Gamma(\alpha - 1)\, \cos \tfrac{1}{2}\pi\alpha \; n^{1-\alpha},$$

with

(7.17) $$A_n = \sum_{j=0}^{\infty} a_{2j+n+1}.$$

When $a_n \downarrow 0$, (7.16) can be replaced by

$$a_n \sim (2/\pi)\, K\, \Gamma(\alpha)\, \cos \tfrac{1}{2}\pi\alpha \; n^{-\alpha}.$$

When $\alpha = 1$, Theorem 7.4 fails. We might look for replacements of two kinds: conditions for $f(x) \to K$ or conditions for $a_n \sim K/n$. The first problem is trivial, since it follows from Theorem 5.34 that $f(x)$ approaches a limit if and only if $\sum a_n$ converges, provided that $a_n \geq 0$. (Consequently the theorems about cosine series in § 5 also fit into the present section.) The second problem is open.

Question 7.18. Find a necessary and sufficient condition for $a_n \sim K/n$ (or for $a_n = O(n)$) when $a_n \downarrow 0$.

Question 7.19. Extend Theorems 7.10 and 7.14 to larger values of α by introducing $g(x) - x\, g'(0)$, $f(x) - f(0) - \tfrac{1}{2} x^2\, f''(0)$, and so on.

If we look for "O-theorems" corresponding to the "o-theorems" 7.10, 7.14, we are led naturally to conditions for g or f to satisfy Lipschitz conditions.

Theorem 7.20[1]). *Let* $\lambda_n \geq 0$ *and let* λ_n *be the sine or cosine coefficients of* φ. *Then* $\varphi \in \mathrm{Lip}\,\gamma\ (0 < \gamma < 1)$ *if and only if*

(7.21) $$\sum_{k=n}^{\infty} \lambda_k = O(n^{-\gamma}),$$

or equivalently

(7.22) $$\sum_{k=1}^{n} k\, \lambda_k = O(n^{1-\gamma}).$$

When $\lambda_k \downarrow 0$, either (7.21) or (7.22) is equivalent to $\lambda_n = O(n^{-1-\gamma})$, so Theorem 7.20 implies Lorentz's theorem[2]):

Theorem 7.23. *If* $\lambda_n \downarrow 0$ *and* λ_n *are the sine or cosine coefficients of* φ, *then* $\varphi \in \mathrm{Lip}\,\gamma\ (0 < \gamma < 1)$ *if and only if* $\lambda_n = O(n^{-1-\gamma})$.

Further theorems on Lipschitz conditions and integrated Lipschitz conditions are given by ALJANČIĆ and TOMIĆ [1] and ALJANČIĆ [1]. The following theorem of ASKEY [1] gives a very general result connecting "smoothness" of a function with its coefficients; it includes Theorem 7.23 and a number of results on integrated Lipschitz conditions (LORENTZ [1], LEINDLER [1], KONYUŠKOV [1]).

---

[1]) BOAS [7].
[2]) LORENTZ [1]; see also BARI [1], pp. 208—210, 678.

Theorem 7.24. *Let* $0 < \alpha < 2$, $1 < p < \infty$, $1 \leq q \leq \infty$, *and let* $a_n \downarrow 0$ *be the cosine coefficients of* f. *Then* $f \in \Lambda(\alpha, p, q)$, *i.e.*

$$\left\{ \int \left[\int \left| \frac{f(x+t) + f(x-t) - 2f(x)}{t^{\alpha}} \right|^{p} dx \right]^{q/p} \frac{dt}{t} \right\}^{1/q} < \infty,$$

if and only if

$$\{ \sum a_n^q n^{q(\alpha + 1 - 1/p) - 1} \}^{1/q} < \infty.$$

REES [1] has given an upper bound for the L^1 integral modulus of continuity of φ when $\lambda_n \downarrow 0$.

Question 7.25. Find theorems on Lipschitz conditions for positive or monotonic functions.

The situation is more complicated when $\gamma = 0$ or 1. By Lip 1 we mean the class of integrals of bounded functions, and by Lip 0 the class of bounded measurable functions. For cosine series and $\gamma = 0$, the relevant theorem is the Paley-Szász theorem 5.34:

Theorem 7.26. *If* $a_n \geq 0$ *then* a_n *are the cosine coefficients of a bounded function, or more generally of a function such that*

$$\int_0^x dt \int_0^t f(u) \, du = O(x^2),$$

if and only if $\sum a_n$ *converges.*

For sine series and $\gamma = 0$, we have the following result[1]).

Theorem 7.27. *If* $b_n \downarrow 0$, *the conditions* $n b_n = O(1)$ *and* $n b_n \to 0$ *are necessary and sufficient, respectively, for* $\sum b_n \sin nx$ *to be boundedly convergent (and hence the Fourier series of a bounded function), and for the series to be uniformly convergent (and hence the Fourier series of a continuous function).*

Compare Theorem 7.37 below.

For sine series and $\gamma = 1$, we have a theorem equivalent to Theorem 7.26[2]).

Theorem 7.28. *If* $b_n \geq 0$ *and* b_n *are the sine coefficients of* g, *then* $g \in$ Lip 1 *if and only if*

(7.29) $\sum k \lambda_k$ *converges.*

Note that (7.29) is (7.22) for $\gamma = 1$; it is not equivalent to (7.21) in this case.

The proofs of Theorems 7.20 and 7.28 show that (7.22) is sufficient for $\varphi \in$ Lip γ and necessary for $|\varphi(x) - \varphi(0)| = O(x^{\gamma})$ as $x \to 0+$. Hence Lipschitz behavior $(0 < \gamma < 1)$ of Fourier series with positive coefficients is propagated from the origin to other points in accordance with the general principle stated in § 1. However, the principle breaks down for Lip 1 and cosine series. Let Λ_* be the class of measurable functions f for which $f(x+2h) + f(x-2h) - 2f(x)$ is uniformly $O(h)$.

---

[1]) See ZYGMUND [2], vol. 1, pp. 182—183.

[2]) For the theorems from here to 7.37 cf. BOAS [7].

Theorem 7.30. *If $a_n \geq 0$ and a_n are the Fourier coefficients of f, then*

$$(7.31) \qquad \sum_{k=1}^{n} k^2 a_k = O(n)$$

or equivalently

$$(7.32) \qquad \sum_{k=n}^{\infty} a_k = O(1/n),$$

if and only if either $f(x) - f(0) = O(x)$ or $f \in \Lambda_*$.

We could replace O's by o's and Λ_* by the class of smooth functions.

Since Lip 1 is the class of functions with essentially bounded derivative, Theorem 7.30 can be given an equivalent form as a theorem on sine series.

Theorem 7.33. *If $b_n \geq 0$ and b_n are the sine coefficients of g, then*

$$(7.34) \qquad \sum_{k=n}^{\infty} k^{-1} b_k = O(1) \qquad\qquad (or\ o(1))$$

if and only if g is $(C, 1)$ bounded or approaches 0 in the $(C, 1)$ sense, i.e.

$$x^{-1} \int_{0}^{x} g(t)\ dt = O(1) \qquad\qquad (or\ o(1)).$$

The Weierstrass function $\sum 2^{-n} \cos 2^n x$ shows that (7.31) or (7.32) is not sufficient for $f \in$ Lip 1.

It is possible to give necessary and sufficient conditions for a cosine series with positive coefficients to belong to Lip 1.

Theorem 7.35. *If $a_n \geq 0$ and a_n are the cosine coefficients of f, then $f \in$ Lip 1 if and only if (7.32) holds and $\sum k\, a_k \sin kx$ has uniformly bounded partial sums. More precisely, $f(x+h) - f(x) = O(h)$ for a given x if and only if $\sum k\, a_k \sin kx$ has its partial sums bounded for that x.*

The corresponding theorem with o instead of O is worth stating separately.

Theorem 7.36. *If $a_n \geq 0$ and a_n are the cosine coefficients of f, and if f is smooth (or equivalently if (7.32) holds) then $f'(x)$ exists if and only if $\sum k\, a_k \sin kx$ converges; f' is continuous if and only if $\sum k\, a_k \sin kx$ converges uniformly.*

This can be restated as a theorem for sine series for $\gamma = 0$.

Theorem 7.37. *If $b_n \geq 0$ and if $\sum b_n \sin nx$ is the formal derivative of the Fourier series of a function G, and*

$$\sum_{k=n}^{\infty} k^{-1} b_k = o(1/n),$$

then $\sum b_n \sin nx$ converges at x if and only if $G'(x)$ exists, and converges uniformly if and only if G' is continuous.

It is interesting to note that if G is continuous then $\sum b_n \sin nx$ converges for uncountably many values of x (under the hypotheses of Theorem 7.37), since a smooth function is differentiable at uncountably many points (ZYGMUND [2], vol. 1, p. 43).

In particular, if $\sum b_n \sin nx$ is the Fourier series of g under the hypotheses of 7.37, it converges almost everywhere, and uniformly if g is continuous. For, in this case $G = \int g$ is smooth and its derivative exists everywhere. We therefore have a localization of Paley's theorem that if the Fourier sine series of a continuous function has positive coefficients, it converges uniformly[1].

We shall prove prove Theorems 7.1 and 7.20; the other theorems of this section can either be proved in similar ways, or deduced from Theorems 7.1 and 7.4 by partial integration and partial summation.

Proof of Theorem 7.1. First let $b_n \downarrow 0$ and $b_n \sim B n^{-\alpha}$, $0 < \alpha < 1$, $B > 0$ (the cases where $\alpha = 1$ or $B = 0$ require modifications in the argument). Then

$$\left| g(x) - B \sum_{n=1}^{\infty} n^{-\alpha} \sin nx \right|$$

$$\leq \left| \sum_{n=1}^{M/x} (b_n - B n^{-\alpha}) \sin nx \right| + \left| \sum_{M/x}^{\infty} b_n \sin nx \right| + \left| \sum_{M/x}^{\infty} B n^{-\alpha} \sin nx \right|$$

$$= S_1 + S_2 + S_3,$$

where M is a number to be chosen later. Since the sums

$$\sum_{n=p}^{q} \sin nx$$

are uniformly $O(1/x)$, and $b_n \downarrow 0$, partial summation shows that S_2 and S_3 are bounded by constant multiples of $M^{-\alpha} x^{\alpha-1}$; hence if, with a given $\varepsilon > 0$, we take M large enough, we have

$$S_2 + S_3 \leq \varepsilon x^{\alpha-1}.$$

Now with M fixed, and N so large that $|b_n - B n^{-\alpha}| < \varepsilon n^{-\alpha}$ for $n \geq N$, we have

$$S_1 \leq \left| \sum_{n=1}^{N} (b_n - B n^{-\alpha}) \sin nx \right| + \varepsilon \left| \sum_{n=N}^{M/x} n^{-\alpha+1} x \right| \leq O(x) + \varepsilon O(x^{\alpha-1}) \leq \varepsilon x^{\alpha-1}$$

if x is small enough. Then (7.2), with the correct K, follows from Lemma 2.13.

Conversely, suppose that (7.2) holds. Then

$$b_n = 2 \pi^{-1} \int g(x) \sin nx \, dx$$

(since the b_n are generalized sine coefficients), and

$$\sum_{k=1}^{n} \left(1 - \frac{k}{n+1} \right) b_k = 2 \pi^{-1} \int g(x) \tilde{K}_n(x) \, dx,$$

---

[1] See HARDY and ROGOSINSKI [2] p. 72, or BARI [1], p. 277.

where $\tilde{K}_n$ is the conjugate Fejér kernel. Hence (since $\tilde{K}_n \geq 0$)

$$\sum_{k=1}^{n} \left(1 - \frac{k}{n+1}\right) b_k = 2\pi^{-1} \int (K + o(1)) x^{\alpha-1} \tilde{K}_n(x) \, dx.$$

Now

$$2\pi^{-1} \int x^{\alpha-1} \tilde{K}_n(x) \, dx$$

are the $(C, 1)$ means of $\sum \beta_n$, where β_n are the sine coefficients of $x^{\alpha-1}$. But

$$\beta_n \sim 2\pi^{-1} \Gamma(\alpha) \sin \tfrac{1}{2}\alpha\pi \, n^{-\alpha},$$

and hence

(7.38) $$\sum_{k=1}^{n} \left(1 - \frac{k}{n+1}\right) b_k \sim 2\pi^{-1} \Gamma(\alpha) \sin \tfrac{1}{2}\alpha\pi \, \frac{n^{1-\alpha}}{(2-\alpha)(1-\alpha)}.$$

From this (7.3) follows by an elementary Tauberian argument, since the b_k decrease.
   In fact, if we write

$$s_n = \sum_{k=1}^{n} b_k,$$

(7.38) says that

$$s_1 + s_2 + \cdots + s_n \sim L n^{2-\alpha},$$

and we have s_n increasing. Taking $\varepsilon > 0$ and $\mu > 1$, we have, for sufficiently large n,

$$\sum_{k=1}^{n\mu+1} s_k < (L+\varepsilon) n^{2-\alpha} \mu^{2-\alpha},$$

$$\sum_{k=1}^{n-1} s_k > (L-\varepsilon) n^{2-\alpha},$$

$$s_n n(\mu-1) \leq \sum_{k=n}^{n\mu+1} s_k < n^{2-\alpha}((L+\varepsilon)\mu^{2-\alpha} - (L-\varepsilon)),$$

and so

$$\limsup_{n \to \infty} n^{\alpha-1} s_n \leq \frac{(L+\varepsilon)\mu^{2-\alpha} - (L-\varepsilon)}{\mu-1}.$$

Letting $\varepsilon \to 0$ and then $\mu \to 1$ yields

$$\limsup_{n \to \infty} n^{\alpha-1} s_n \leq (2-\alpha) L.$$

Similarly

$$\liminf_{n \to \infty} n^{\alpha-1} s_n \geq (2-\alpha) L,$$

and thus

$$s_n \sim (2-\alpha) L n^{1-\alpha}.$$

Repeating the argument with b_n instead of s_n, and allowing for the fact that b_n decreases whereas s_n increases, we get

$$b_n \sim (2-\alpha)(1-\alpha) L n^{-\alpha},$$

and this is (7.3).

We now prove Theorem 7.20; the weaker conclusion (as compared with Theorems 7.1 and 7.4) allows us to make the proof much simpler.

We need two lemmas.

Lemma 7.39. *Let $\mu_k \geq 0$, $\delta > \beta > 0$. Then*

$$\sum_{k=1}^{n} k^{\delta} \mu_k = O(n^{\beta})$$

is equivalent to

$$\sum_{k=n}^{\infty} \mu_k = O(n^{\beta-\delta}).$$

This is a straightforward application of partial summation.

Lemma 7.40. *If $\mu_k \geq 0$, $\sum \mu_k$ converges, and $0 < \beta < 2$, then*

(7.41)
$$\sum_{k=1}^{\infty} \mu_k(1 - \cos kx) = O(x^{\beta})$$

if and only if

(7.42)
$$\sum_{k=n}^{\infty} \mu_k = O(n^{-\beta}).$$

Let (7.41) hold. Then

$$\sum_{k=1}^{1/x} k^2 \mu_k \frac{1 - \cos kx}{k^2 x^2} \leq \sum_{k=1}^{\infty} k^2 \mu_k \frac{1 - \cos kx}{k^2 x^2} = O(x^{\beta-2}).$$

Since $t^{-2}(1 - \cos t)$ decreases on $(0, 1)$, it follows that

(7.43)
$$\sum_{k=1}^{1/x} k^2 \mu_k = O(x^{\beta-2}),$$

and (7.42) follows from Lemma 7.39.

Conversely let (7.42) hold. Then (7.43) holds, by Lemma 7.39. Hence

$$\sum_{k=1}^{\infty} \mu_k(1 - \cos kx) \leq \sum_{1}^{1/x} + \sum_{1/x}^{\infty} \leq O(1) x^2 \sum_{k=1}^{1/x} k^2 \mu_k + O(1) \sum_{k=1/x}^{\infty} \mu_k = O(x^{\beta}).$$

Now let $0 < \gamma < 1$ and let $\sum a_n \cos nx$ be the Fourier series of f, with $a_n \geq 0$. First let $f(x) - f(0) = O(x^{\gamma})$. Then the Fourier series of f converges at $x = 0$ (by Dini's test) and we have

(7.44)
$$\sum a_k(1 - \cos kx) = O(x^{\gamma}).$$

By Lemma 7.40 this is equivalent to (7.21), i.e.

$$\sum_{k=n}^{\infty} a_k = O(n^{-\gamma}),$$

and by Lemma 7.39 this in turn is equivalent to (7.22).

Conversely, let (7.21) hold. We have

$$\begin{aligned}
|f(x+2h) - f(x)| &= |\sum a_k [\cos k(x+2h) - \cos kx]| \\
&= 2|\sum a_k \sin k(x+h) \sin kh| \\
&\leq 2 \sum_{k=1}^{1/h} a_k \sin kh + 2 \sum_{k=1/h}^{\infty} a_k |\sin kh| \\
&\leq 2h \sum_{k=1}^{1/h} k a_k + 2 \sum_{k=1/h}^{\infty} a_k.
\end{aligned}$$

The second sum is $O(h^\gamma)$ by hypothesis and the first sum is $O(h^\gamma)$ by Lemma 7.39. This shows that $f \in \text{Lip } \gamma$.

The proof for sine series is similar.

§ 8. More general classes of functions; conditional convergence

In the problems with which we started in § 1, we had a basic class of functions φ (decreasing, or positive, or with decreasing or positive Fourier coefficients) and multipliers $t^{-\gamma}$ and $n^{\gamma-1}$; the problem was to show that $\int |\varphi(t)|\, t^{-\gamma}\, dt$ exists if and only if $\sum n^{\gamma-1} |\lambda_n|$ is finite, or at least that the existence of one of these implies the existence of the other. In looking for generalizations we could, for example, (A) keep the same basic class for φ and $\{\lambda_n\}$ but use a class of multipliers more general than powers; (B) widen the basic class; (C) widen the basic class but show only that absolute convergence of integral or series implies conditional convergence of series or integral; (D) widen the basic class and show that conditional convergence of series or integral implies conditional convergence of the other; (E) generalize (A) or (B) to L^p, $p > 1$. A number of examples of such generalizations have already turned up in a natural way in the preceding sections. Here we mention a number of others, and refer to the literature for additional results.

(A) A natural class of generalized powers consists of functions $x^{-\gamma} L(1/x)$, where L is slowly increasing (i.e., $L(c\,x)/L(x) \to 1$).

Theorem 8.1[1]). If $0 < \gamma < 1$ (*for cosine series*) or $0 < \gamma < 2$ (*for sine series*) and $\lambda_n \downarrow 0$ then $\sum n^{\gamma-1} L(n)\, \lambda_n$ converges if and only if $x^{-\gamma} L(1/x)\, \varphi(x) \in L$.

There is also an extension to $1 < \gamma < 3$ for cosine series.
However, a much wider class of multipliers can be used.

Theorem 8.2[2]). If $b_n \downarrow 0$, if $G(x) \geq 0$ and $x\, G(x) \in L$, and

$$(8.3) \qquad \sum n\, b_n \int_0^{1/n} x\, G(x)\, dx < \infty$$

then $g(x)\, G(x) \in L$; the converse holds if $x^{-1} G(x) \in L$ (or if G satisfies some more general but more complicated conditions). If $a_n \downarrow 0$, $F(x) \geq 0$, $F(x) \in L$, and

$$\sum a_n \int_0^{1/n} F(x)\, dx < \infty$$

then $F\, f \in L$; the converse holds if

$$\int_t^1 x^{-1} F(x)\, dx \leq C\, F(t)$$

(or if F satisfies some more complicated conditions).

---

[1]) ALJANČIĆ, BOJANIĆ and TOMIĆ [1].
[2]) PEYERIMHOFF [1] for sine series, YANG [1] for cosine series. See also CHEN [4], where Theorems 4.21 and 4.22 are also generalized.

ADAMOVIĆ [1] has given a similar result with decreasing functions g instead of decreasing coefficients b_n.

The numbers (8.3) are closely related to the sine coefficients B_n of G, since

$$\tfrac{1}{2}\pi B_n = \int_0^{1/n} G(x)\sin nx\,dx + \int_{1/n}^{\pi} G(x)\sin nx\,dx = \beta_n' + \beta_n'',$$

say; β_n' is, when $G(x) \geq 0$, nearly

$$n \int_0^{1/n} x\,G(x)\,d\,x,$$

and $\sum b_n\,\beta_n''$ converges if $G(x)$ satisfies mild additional conditions. Hence the following theorems[1]) are closely related to Theorem 8.2.

Theorem 8.4. *If $b_n\downarrow 0$, if $G(x) \geq 0$ and $x\,G(x)\in L$, and B_n are the generalized sine coefficients of G, then $g(x)\,G(x)\in L$ if $\sum b_n\,B_n$ converges or more generally if*

$$\liminf_{N\to\infty} \sum_{n=1}^{N} b_n\,B_n < \infty,$$

and Parseval's theorem holds if $\sum b_n\,B_n$ converges. The converse holds if $G(x)$ decreases.

This includes Theorem 4.1 (take $G(x) = x^{-\gamma}$), and Theorem 3.8 (take g in Theorem 3.8 to be G here, and take $g(x)$ here to be $x^{-\gamma}$).

Theorem 8.5. *If $a_n\downarrow 0$ and $F(x)$ decreases then $\sum a_n\,A_n$ converges if and only if*

$$\int_{0+}^{\pi} f(x)\,F(x)\,dx$$

converges, and Parseval's theorem holds if either side is finite.

Question 8.6. Is there an analogue of Theorem 8.4 with $b_n \geq 0$ instead of $G(x) \geq 0$?

Theorems 8.4 and 8.5 deal with the basic class of functions that have decreasing Fourier coefficients, and with decreasing or positive multipliers. The next two theorems deal, respectively, with the class of decreasing functions, with decreasing multiplier functions; and with the class of functions that have decreasing coefficients, with decreasing multiplier coefficients.

Theorem 8.7[2]). *If f and F decrease and are integrable then $\sum a_n\,A_n$ converges (in general, conditionally) if and only if $fF \in L$, and Parseval's formula holds. If g and G are positive and decreasing, and $x\,G(x)$ and $x\,g(x)\in L$, then $\sum b_n\,B_n$ converges (absolutely, since b_n and B_n are positive) if and only if $gG \in L$, and Parseval's formula holds.*

---

[1]) EDMONDS [2], Theorems 27 and 19 (with a slight generalization of 19 communicated to me by EDMONDS).

[2]) EDMONDS [2], Theorems 15 and 17 (slightly extended).

This contains Theorem 3.8.

Theorem 8.8[1]). *If a_n and $A_n \downarrow 0$ and $\sum a_n A_n$ converges then*

$$\int_{0+}^{\pi} f(x) F(x) \, dx$$

converges, and Parseval's formula holds. If b_n and $B_n \downarrow 0$ then $\sum b_n B_n$ converges if and only if $gG \in L$, and Parseval's formula holds.

This theorem includes Theorem 4.1.

Edmonds proves many other theorems and discusses the roles of the various hypotheses in detail. Her papers should be required reading for anyone interested in the material of the present section.

In the original case of power multipliers we obtained absolute-convergence theorems for cosine series with decreasing coefficients from theorems for sine series with positive coefficients by integrating or summing by parts. This suggests the following problem.

Question 8.9. Investigate Parseval's theorem for the class of odd functions with positive coefficients.

(B) In many of our theorems the basic class of functions has been the class with decreasing or positive coefficients. Several of these theorems have been extended to the case where the coefficients are quasimonotonic or quasipositive.

A sequence $\{\lambda_n\}$ is called quasimonotonic if $n^{-\beta} \lambda_n \downarrow 0$ for some β, or equivalently if $\Delta \lambda_n \geq -\alpha n^{-1} \lambda_n$ for some α; and quasipositive if λ_n are the differences of a quasimonotonic sequence. More generally, $\{\lambda_n\}$ is δ-quasimonotonic if $\lambda_n \to 0$, $\lambda_n > 0$ ultimately, and $\Delta \lambda_n \geq -\delta_n$. If $\delta_n = \alpha \lambda_n/n$, a δ-quasimonotonic sequence is quasimonotonic[2]).

Theorems 4.1, 4.2 remain true when "decreasing" is replaced by "quasimonotonic"[3]); also when it is replaced by "δ-quasimonotonic" provided that $\sum n^\gamma \delta_n < \infty$. Theorems 4.8 and 4.9 remain true when "positive" is replaced by "δ-quasipositive", provided that $\sum n^{\gamma-1} \delta_n < \infty$. Theorem 6.3 remains true when $\{\lambda_n\}$ is quasimonotonic[4]), and has a number of applications when so generalized. Presumably most of the theorems with decreasing coefficients can be generalized similarly. Those with monotonic functions can be generalized with an appropriate definition of a quasimonotonic function whose domain is $(0, \pi)$. In particular, Theorem 6.3 remains true with quasimonotonic functions instead of quasimonotonic coefficients, and can be proved by the method of ASKEY and WAINGER [1][5]). Theorem 8.1 holds with quasimonotonic coefficients (YONG 1]).

---

[1]) EDMONDS [2], Theorems 13 and 16.

[2]) For references see BOAS [5]. Theorem 2 of that paper with $\gamma = 0$ should have $\sum \delta_n \log n < \infty$ (correction pointed out by C. C. GANSER).

[3]) O'SHEA [1], SHAH [1].

[4]) ASKEY and WAINGER [1], YONG [1].

[5]) Communication from R. ASKEY.

(C) In the first theorem here we assume very little about φ or $\{\lambda_n\}$, but use power functions as multipliers[1]).

Theorem 8.10. *If $0 < \gamma \leq 1$ for cosine series or $-1 < \gamma \leq 1$ for sine series, if λ_n are the cosine or (generalized) sine coefficients of φ, and $x^{-\gamma}\varphi(x) \in L$ then $\sum \lambda_n n^{\gamma-1}$ converges. If $0 < \gamma < 1$ for cosine series or $0 \leq \gamma < 1$ for sine series, and $\sum |\lambda_n| n^{\gamma-1}$ converges, then*

$$\int_{0+}^{\pi} x^{-\gamma}\varphi(x)\,dx$$

converges.

We do not necessarily have absolute convergence in the conclusions (HEYWOOD [4]). For $0 < \gamma < 1$, Theorem 8.10 appeared incidentally in § 4 as Theorem 4.13. Compare Theorems 8.5 and 8.8.

Theorem 8.10 can be extended by replacing the power-function multipliers by something more general that, in particular, covers powers multiplied by slowly increasing functions.

Theorem 8.11[2]). *Let $\varphi \in L$ and let*

$$\sum |\lambda_n| \int_0^{1/n} \psi(x)\,dx$$

converge, where ψ is positive and decreasing; then

$$\int_{0+}^{\pi} \varphi(x)\,\psi(x)\,dx$$

converges and Parseval's formula holds for φ and ψ. Conversely let $\mu_n \downarrow 0$,

$$M(u) = \sum_{n \leq u} \mu_n,$$

and $\varphi(x) M(1/x) \in L$; then $\sum \lambda_n \mu_n$ converges and Parseval's theorem holds.

For sine series the second part holds more generally with $M(u)$ replaced by

$$M_1(u) = u^{-1} \sum_{n \leq \alpha} n \mu_n,$$

and the first part holds with

$$\int_0^{1/n} \psi(x)\,dx$$

replaced by

$$n \int_0^{1/n} x\,\psi(x)\,dx.$$

Theorem 11 includes some results of ROBERTSON [3]. There are similar results by CHEN [5].

---

[1]) IZUMI and SATÔ [1], HEYWOOD [4].
[2]) M. and S. IZUMI [1].

(D) HEYWOOD [5], [6] has proved the following theorems with conditional convergence in both hypothesis and conclusion.

Theorem 8.12. *Let* $\sum |b_n|$ *converge and let* $1 < \gamma < 2$. *If* $\sum n^{\gamma-1} b_n$ *converges then*

$$\int_{0+}^{1} x^{-\gamma} g(x)\, dx$$

converges. Conversely if this integral converges and $b_n > -C n^{-\gamma}$ *then* $\sum n^{\gamma-1} b_n$ *converges.*

Theorem 8.13. *Let* $x\, g(x) \in L$. *If* $-1 < \gamma < 0$ *or if* $0 < \gamma < 1$ *and* $g(x) > -C x^{\gamma-1}$ *then* $\sum n^{\gamma-1} b_n$ *converges if*

$$\int_{0+}^{\pi} x^{-\gamma} g(x)\, dx$$

converges. If $-1 < \gamma < 1$ *and* $g(x) > -C x^{\gamma-1}$ *then the integral converges if the series converges.*

In the next theorem the multiplier functions are more general but an order condition is imposed on $\varphi(x)$ at 0.

Theorem 8.14[1]). *Let* $\varphi \in L$ *and suppose that*

$$M(n) \int_{0}^{\mu_n} |\varphi(x)|\ dx \to 0, \quad n \to \infty,$$

where M is defined in Theorem 8.11. Then $\sum \lambda_n \mu_n$ *converges if and only if*

$$\int_{0+}^{\pi} \varphi(x)\, \psi(x)\, dx$$

converges.

(E) Theorems generalizing those of § 6 for L^p classes, with $x^{-\gamma}$ replaced by $x^{-\gamma} L(1/x)$, L slowly increasing, have been given by ADAMOVIĆ [1], IGARI [1] and YONG [2].

A different kind of generalization has been discussed in great detail by CHEN [2], [3], who generalizes not only the power-function multipliers but also the L^p classes. We quote a specimen result.

Theorem 8.15. *Let* Φ *be a positive function such that* $x^{-1} \Phi(x)$ *increases and* $x^{-k} \Phi(x)$ *decreases for some* $k > 1$; *let* Ψ *be strictly increasing with* $x^{\delta-1} \Psi(x)$ *strictly decreasing for some* $\delta > 0$. *If* $b_n \downarrow 0$ *then* $\Phi(|g(x)|)/\Psi(x) \in L$ *if and only if*

$$\sum \Phi(n\, b_n)\, n^{-2}/\Psi(1/n) < \infty.$$

---

[1]) M. and S. IZUMI [1].

§ 9. Trigonometric integrals

We may expect theorems for trigonometric integrals to be similar to theorems for trigonometric series and to be provable by similar methods. They are likely to be different in detail. Also if, for example,

$$f(x) = \int_0^\infty a(t) \cos xt \, dt,$$

we can discuss the integrability of $x^{\gamma-1} a(x)$ and of $x^{-\gamma} f(x)$ in neighborhoods either of 0 or of ∞. On the other hand, since

$$a(t) = 2\pi^{-1} \int_0^\infty f(x) \cos xt \, dx,$$

there is no real distinction between, for example, theorems with decreasing a and theorems with decreasing f.

The following theorems are analogues of some of the theorems of § 4. They were proved by Sz.-Nagy [1]; those for sine transforms are deducible from theorems of Edmonds [2]. To bring out the analogy with the series theorems we let f and g be the cosine and sine transforms of a and b.

Theorem 9.1. *If $0 \le \gamma < 1$, $b(t) \downarrow 0$ on $(0, \infty)$, and $t \, b(t) \in L$ on every finite interval, then $g(x) \, x^{-\gamma} \in L(0, 1)$ or $L(1, \infty)$ if and only if $t^{\gamma-1} b(t) \in L(1, \infty)$ or $L(0, 1)$, respectively.*

Theorem 9.2. *If $0 < \gamma < 1$, $a(t) \downarrow 0$ on $(0, \infty)$, and $a(t) \in L$ on every finite interval, then $f(x) \, x^{-\gamma} \in L(0, 1)$ or $L(1, \infty)$ if and only if $t^{\gamma-1} a(t) \in L(1, \infty)$ or $L(0, 1)$, respectively.*

Theorem 9.3. *If $a(t)$ is monotonic near 0 and ∞, of bounded variation on every $(\alpha, \beta) \, (0 < \alpha < \beta < \infty)$, $a(\infty) = 0$, and $a \in L(0, \infty)$, then $f(x)/x \in L(0, 1)$ or $L(1, \infty)$, respectively, if and only if $a(t) \log t \in L(1, \infty)$ and*

$$\int_0^\infty a(t) \, dt = 0,$$

or $a(t) \log t \in L(0, 1)$.

The general results of Edmonds [1], [2] verify various cases of the principle that if we have a pair of functions and their Fourier transforms, then Parseval's formula holds if two (not transforms of each other) of the four functions entering the formula are monotonic and one of the two integrals converges. From the large number of theorems proved by Edmonds, I select the following as particularly relevant to the theme of the present monograph[1]) and recommend that a reader who is interested in theorems of this kind should consult her papers.

We shall say that $b(x)$ is the sine transform of $g(t)$ if either $tg(t) \in L(0, 1)$ and $g(t) \in L(1, \infty)$, and

(9.4)
$$b(x) = (2/\pi)^{1/2} \int_0^\infty g(t) \sin xt \, dt,$$

or $g(t)$ decreases to 0 on $(0, \infty)$ and the integral in (9.4) is interpreted as a Cauchy limit (at both ends). Similarly B is the sine transform of G; and a, A are the cosine transforms of f, F.

Theorem 9.5. *If g and G decrease to* 0 *on* $(0, \infty)$ *and* $x g(x)$, $x G(x) \in L(0, 1)$, *then* $bB \in L(0, \infty)$ *if and only if* $gG \in L$ (*and then Parseval's formula holds*).

In the next theorem the conclusion is the same but instead of having G monotonic we assume B monotonic and define G as the sine transform of B.

Theorem 9.6. *If g and B decrease to* 0 *on* $(0, \infty)$ *and* $x g(x)$, $x B(x) \in L(0, 1)$, *then* $bB \in L(0, \infty)$ *if and only if* $gG \in L$ (*and then Parseval's formula holds*).

Theorem 9.1 can be deduced from either of these theorems, since if $0 < \gamma < 1$, the sine transform of $t^{\gamma - 1}$ is a constant multiple of $x^{-\gamma}$, so in the situation of Theorem 9.1 we have both a monotonic function and a second monotonic function with a monotonic transform.

The situation for cosine transforms is more complex, unless we use convex functions instead of decreasing functions, since a positive decreasing function has a positive sine transform but not necessarily a positive cosine transform. We quote the following two theorems (EDMONDS [2], Theorems 1 and 6), and an unsolved problem (Edmonds' problem B).

Theorem 9.7. *If f and $F \downarrow 0$ on* $(0, \infty)$ *and* $fF \in L$, *then*

$$\int_{0+}^{\to \infty} a(x) A(x) \, dx = \int_0^\infty f(t) F(t) \, dt,$$

where the integral on the left is a Cauchy limit at both ends.

Theorem 9.8. *If f and $A \downarrow 0$ on* $(0, \infty)$ *then*

$$\int_{0+}^{\to \infty} a(x) A(x) \, dx = \int_{0+}^{\to \infty} f(t) F(t) \, dt$$

provided either side exists.

---

[1]) I have incorporated slight modifications, suggested by EDMONDS, into Theorems 9.5 and 9.6 (her Theorems 2 and 4).

Question 9.9. If f and $F \downarrow 0$ on $(0, \infty)$ and

$$\int_{0+}^{\to \infty} a(x)\, A(x)\, dx$$

exists, does

$$\int_0^\infty f(t)\, F(t)\, dt$$

exist in some sense; and if so, does Parseval's formula hold?

We note that the integral analogues of Theorems 6.3 and 6.6 are valid, and can be proved, for example, by the method of ASKEY and WAINGER [1].

The characteristic functions of distribution functions (in the terminology used in the theory of probability) have the form

$$\varphi(x) = \int_{-\infty}^{\infty} e^{ixt}\, d\, F(t),$$

where F is an increasing function with $F(-\infty) = 0$, $F(\infty) = 1$. These are generalizations of $\sum a_n \cos n x$ with $a_n \geq 0$ and $\sum a_n < \infty$. There are corresponding generalizations of Theorems 7.20 and 4.4 (BOAS [8]).

Theorem 9.10. *If* $0 < \gamma < 1$ *then* $\varphi \in \mathrm{Lip}\, \gamma$ *if and only if*

$$F(x) - F(\pm \infty) = O(|x|^{-\gamma}), \quad |x| \to \infty.$$

There are also theorems for $\gamma = 1$.

Theorem 9.11. *If* $1 < \gamma < 2$, *then*

$$\int_a^{a+1} (x-a)^{-\gamma} |\varphi(x) - \varphi(a)|\, d x < \infty$$

for every real a, if and only if

$$\int_{-\infty}^{\infty} |t|^{\gamma - 1}\, d\, F(t)$$

converges.

Further theorems on integrals are given, for example, by ROBERTSON [3] and HEYWOOD [5].

Bibliography

ADAMOVIĆ, D.: [1] Généralisations de deux théorèmes de Zygmund—B.Sz.-Nagy. Acad. Serbe Sci. Publ. Inst. Math. **12**, 81—100 (1958).

ALJANČIĆ,S.:[1] On the integral moduli of continuity in L^p ($1 < p < \infty$) of Fourier series with monotone coefficients. Proc. Amer. Math. Soc. **17**, 287—294 (1966).

ALJANČIĆ, S., and M. TOMIĆ: [1] Über den Stetigkeitsmodul von Fourier-Reihen mit monotonen Koeffizienten. Math. Z. **88**, 274—284 (1965).

ALJANČIĆ, S., R. BOJANIĆ and M. TOMIĆ:[1] Sur l'intégrabilité de certaines séries trigono-métriques. Acad. Serbe Sci. Publ. Inst. Math. **8**, 67—84 (1955).

— [2] Sur le comportement asymptotique au voisinage de zéro des séries trigonométriques de sinus à coefficients monotones. Acad. Serbe Sci. Publ. Inst. Math. **10**, 101—120 (1956).

ASKEY, R.:[1] Smoothness conditions for Fourier series with monotone coefficients. Acta Sci. Math. (Szeged) (to appear).

— [2] A transplantation theorem for Jacobi coefficients. Pacific J. Math. (to appear).

ASKEY, R., and R. P. BOAS, Jr.: [1] Fourier coefficients of positive functions (to appear).

ASKEY, R., and S. WAINGER: [1] Integrability theorems for Fourier series. Duke Math. J. **33**, 223—228 (1966).

— [2] A transplantation theorem between ultraspherical series. Ill. J. Math. **10**, 322—344 (1966).

— [3] A transplantation theorem for ultraspherical coefficients. Pacific J. Math. **16**, 393—405 (1966).

BARI, N. K.: [1] Trigonometric series (in Russian). Gosudarstv. Izdat. Fiz.-Mat. Lit., Moscow, 1961.

BOAS, R. P., Jr.: [1] Integrability of trigonometric series. I. Duke Math. J. **18**, 787—793 (1951).

— [2] Integrability of trigonometric series. II. Quart. J. Math. Oxford Ser. (2) **3**, 217—221 (1952).

— [3] Absolute convergence and integrability of trigonometric series. J. Rational Mech. Anal. **5**, 621—632 (1956).

— [4] Integrability of nonnegative trigonometric series. I, II. Tôhoku Math. J. (2) **14**, 363—368 (1962); **16**, 368—373 (1964).

— [5] Quasi-positive sequences and trigonometric series. Proc. London Math. Soc. (3) **14 A**, 38—46 (1965).

— [6] Asymptotic formulas for trigonometric series. J. Indian Math. Soc. (to appear).

— [7] Fourier series with positive coefficients. J. Math. Anal. **17**, 463—483 (1967).

— [8] Lipschitz behavior and integrability of characteristic functions. Ann. Math. Statist. **38**, 32—36 (1967).

BOSANQUET, L. S., and A. C. OFFORD.: [1] Note on Fourier series. Compositio Math. **1**, 180—187 (1934).

CHEN YUNG-MING: [1] On the integrability of functions defined by trigonometrical series. Math. Z. **66**, 9—12 (1956).

— [2] Some asymptotic properties of Fourier constants and integrability theorems. Math. Z. **68**, 227—244 (1957).

— [3] Some further asymptotic properties of Fourier constants. Math. Z. **69**, 105—120 (1958).

— [4] Some integrability theorems. Proc. Glasgow Math. Assoc. **7**, 101—108 (1965).

— [5] Integrability theorems of Fourier series. Arch. Math. 11, 101—103 (1960).

CHURCH, R. F.: [1] On a constant in the theory of trigonometric series. Math. of Comp. **19**, 501 (1965).

DIKAREV, V. A., and V. I. MACAEV: [1] An exact interpolation theorem (in Russian). Dokl. Akad. Nauk. SSSR **168**, 986—988 (1966).

EDMONDS, S. M.: [1] On the Parseval formulae for Fourier transforms. Proc. Cambridge Philos. Soc. **38**, 1—19 (1942).

— [2] The Parseval formulae for monotonic functions. I, II, III, IV. Proc. Cambridge Philos. Soc. **43**, 289—306 (1947); **46**, 231—248, 249—267 (1950); **49**, 218—229 (1953).

FEJÉR, L.: [1] Trigonometrische Reihen und Potenzreihen mit mehrfach monotoner Koeffizientenfolge. Trans. Amer. Math. Soc. **39**, 18—59 (1936).

GANSER, C. C.: [1] Integrability of ultraspherical series. Duke Math. J. **33**, 539—545 (1966).

GOES, G.: [1] Some spaces of Fourier coefficients. Notices Amer. Math. Soc. **9**, 112 (1962).

GOLUBOV, B. I.: [1] On Fourier series of continuous functions with respect to the Haar system (in Russian). Izv. Akad. Nauk SSSR. Ser. Mat. **28**, 1271—1296 (1964).

GONZÁLEZ-FERNÁNDEZ, J. M.: [1] Integrability of trigonometric series. Proc. Amer. Math. Soc. **9**, 315—319 (1958).

HARDY, G. H.: [1] A theorem concerning Taylor's series. Quart. J. Pure Appl. Math. **44**, 147—160 (1913).

— [2] A theorem concerning trigonometrical series. J. London Math. Soc. **3**, 12—13 (1928).

— [3] Some theorems concerning trigonometrical series of a special type. Proc. London Math. Soc. (2) **32**, 441—448 (1931).

HARDY, G. H., and J. E. LITTLEWOOD: [1] Two theorems concerning Fourier series. J. London Math. Soc. **1**, 19—25 (1926).

HARDY, G. H., and W. W. ROGOSINSKI: [1] Notes on Fourier series (I): On sine series with positive coefficients. J. London Math. Soc. **18**, 50—57 (1943).

— [2] Fourier series. Cambridge: At the University Press, 1944; 2d ed., 1950.

— [3] Notes on Fourier series (III): Asymptotic formulae for the sums of certain trigonometrical series. Quart. J. Math. Oxford Ser. **16**, 49—58 (1945).

HARDY, G. H., J. E. LITTLEWOOD and G. PÓLYA: [1] Inequalities. Cambridge: At the University Press, 1934.

HARTMAN, P., and A. WINTNER: [1] On sine series with monotone coefficients. J. London Math. Soc. **28**, 102—104 (1953).

HEYWOOD, P.: [1] On the integrability of functions defined by trigonometric series. I. Quart. J. Math. Oxford Ser. (2) **5**, 71—76 (1954).

— [2] A note on a theorem of Hardy on trigonometrical series. J. London Math. Soc. **29**, 374—378 (1954).

— [3] On the integrability of functions defined by trigonometric series. II. Quart. J. Math. Oxford Ser. (2) **6**, 77—79 (1955).

— [4] Integrability theorems for trigonometric series. Quart J. Math. Oxford Ser. (2) **13**, 172—180 (1962).

— [5] Some Tauberian theorems for trigonometric series and integrals. Proc. London Math. Soc. (3) (to appear).

— [6] Some theorems on trigonometric series. Quart. J. Math. Oxford Ser. (2) (to appear).

HYLTÉN-CAVALLIUS, C.: [1] Geometrical methods applied to trigonometrical sums. Kungl. Fysiografiska Sällskapets i Lund Förhandlingar [Proc. Roy. Physiog. Soc. Lund] **21**, no. 1 (1950).

IGARI, S.: [1] Some integrability theorems of trigonometric series and monotone decreasing functions. Tôhoku Math. J. (2) **12**, 139—146 (1960).

IZUMI, M., and S. IZUMI: [1] Integrability theorem for Fourier series and Parseval equation. J. Math. Anal. Appl. (to appear).

IZUMI, S.: [1] Some trigonometrical series. III. J. Math. (Tokyo) 1, 128—136 (1953).

— [2] Some trigonometrical series. XI. Tôhoku Math. J. (2) 6, 73—77 (1954).

IZUMI, S., and M. SATÔ: [1] Integrability of trigonometric series. I. Tohoku Math. J. (2) 5, 258—263 (1954).

KONYUŠKOV, A. A.: [1] On Lipschitz classes (in Russian). Izv. Akad. Nauk SSSR. Ser. Mat. 21, 423—448 (1957).

LEINDLER, L.: [1] Über Strukturbedingungen für Fourierreihen. Math. Z. 88, 418—431 (1965).

— [2] Nicht verbesserbare Strukturbedingungen. Studia Math. 26, 155—163 (1966).

LORENTZ, G. G.: [1] Fourier-Koeffizienten und Funktionenklassen. Math. Z. 51, 135—149 (1948).

— [2] Bernstein polynomials. University of Toronto Press, 1953.

LUKE, Y. L., W. FAIR, G. COOMBS, and R. MORAN: [1] On a constant in the theory of trigonometric series. Math. of Comp. 19, 501—502 (1965).

O'SHEA, S.: [1] Note on an integrability theorem for sine series. Quart. J. Math. Oxford Ser. (2) 8, 279—281 (1957).

PAK, I. N.: [1] On properties of the sums of some trigonometric series (in Russian). Dokl. Akad. Nauk SSSR 151, 38—41 (1963).

— [2] Some properties of the sums of special trigonometric series (in Russian). Izv. Vysš. Učebn. Zaved. Matematika No. 6 (49), 113—123 (1965).

PEYERIMHOFF, A.: [1] Über trigonometrische Reihen mit monoton fallenden Koeffizienten. Arch. Math. 9, 75—81 (1958).

REES, C. S.: [1] A bound for the integral modulus of continuity, J. Math. Anal. Appl. (to appear).

ROBERTSON, M. M.: [1] Integrability of trigonometric series. Math. Z. 83, 119—122 (1964).

— [2] Integrability of trigonometric series. II. Math. Z. 85, 62—67 (1964).

— [3] Integrability theorems for trigonometric series and transforms. Math. Z. 91, 20—29 (1966).

SHAH, S. M.: [1] Trigonometric series with quasi-monotone coefficients. Proc. Amer. Math. Soc. 13, 266—273 (1962).

ŠNEĬDER, A. A.: [1] On series of Walsh functions with monotonic coefficients (in Russian). Izv. Akad. Nauk SSSR. Ser. Mat. 12, 179—192 (1948).

STEČKIN, S. B.: [1] On power series and trigonometric series with monotone coefficients (in Russian). Uspehi Mat. Nauk 18, no. 1 (109), 173—180 (1963).

SUNOUCHI, G.: [1] Integrability of trigonometrical series. J. Math. (Tokyo) 1, 99—103 (1953).

SZÁSZ, O.: [1] On the partial sums of certain Fourier series. Amer. J. Math. 59, 696—708 (1937).

S. SZIDON: [1] Reihentheoretische Sätze und ihre Anwendungen in der Theorie der Fourierschen Reihen. Math. Z. 10, 121—127 (1921).

SZ.-NAGY, B.: [1] Séries et intégrales de Fourier des fonctions monotones non bornées. Acta Sci. Math. (Szeged) 13, 118—135 (1949).

TELJAKOVSKIĬ, S. A.: [1] Some estimates for trigonometric series with quasiconvex coefficients (in Russian). Mat. Sb. 63 (105), 426—444 (1964).

TOMIĆ, M.: [1] Einige Sätze über die Positivität der trigonometrischen Polynome. Acad. Serbe Sci. Publ. Inst. Math. 4, 145—146 (1952).

— [2] Sur les sommes trigonométriques à coefficients monotones (in Serbo-Croatian). Srpska Akad. Nauka. Zbornik Radova 18, Matematički Inst. 2, 13—52 (1952).

TURÁN, P.: [1] Über die Partialsummen der Fourierreihe. J. London Math. Soc. **13**, 278—282 (1938).
— [2] On a trigonometrical sum. Ann. Soc. Polon. Math. **25**, 155—161 (1953).
WAINGER, S.: [1] Special trigonometric series in k-dimensions. Mem. Amer. Math. Soc., no. **59** (1965).
WEISS, M. C.: [1] The law of the iterated logarithm for lacunary series and its application to the Hardy-Littlewood series. Dissertation, University of Chicago, 1957.
YANG CHAO-HUI: [1] On the integrability of functions defined by cosine series with monotone decreasing coefficients. Acad. Serbe Sci. Publ. Inst. Math. **12**, 73—80 (1958).
YONG CHI-HSING: [1] Integrability theorems of certain trigonometric series with quasi-monotone coefficients, Monatsh. Math. **69**, 441—450 (1965).
— [2] Integrability theorems of powers of trigonometric series with quasi-monotone coefficients. Math. Z. **91**, 280—293 (1966).
YOUNG, W. H.: [1] On the Fourier series of bounded functions. Proc. London Math. Soc. (1) **12**, 41—70 (1913).
ZYGMUND, A.: [1] Sur les fonctions conjuguées. Fund. Math. **13**, 284—303 (1929).
— [2] Trigonometric series. 2d ed. Cambridge: At the University Press, 1959.

Addendum

KOKILAŠVILI, V. M.: [1] On best approximations by Walsh polynomials and the Walsh-Fourier coefficients (inRussian). Bull. Acad. Polon. Sci. Sér. Sci. Math. Astronom. Phys. **13**, 405—410 (1965).

Index

ADAMOVIĆ 53, 56
ALJANČIĆ 8, 43, 45, 52
ASKEY 6, 13, 19, 35, 36, 37, 45, 54, 59
Asymptotic formulas 5, §7

BARI 22, 45, 48
BOJANIĆ 8, 43, 52
BOSANQUET 33
Bounded variation 14, 15, 23, 27, 28

CESARI 30
Characteristic function 59
CHEN 35, 52, 55, 56
CHURCH 10
Conditional convergence 14, 28, 31, 33, 34, §8
Convergence in L 30
Convex function 13
Convex sequence 27, 30
COOMBS 10
Cosine transforms §9

Decreasing coefficients 3, 5, §§4–7, 52, 53
Decreasing functions 3, 5, §3, 26, 29, 30, 35, 44, 53, §9
DIKAREV 36
Distribution function 59

EDMONDS 13, 14, 53, 54, 57, 58

FAIR 10
FEJÉR 10
Fourier coefficients, generalized 1, 4
Fourier-Stieltjes coefficients 14, 15, 29

GANSER 6, 54
Generalized sine or cosine coefficients 1, 4

GOES 29
GOLUBOV 6
GONZÁLEZ-FERNÁNDEZ 25

Haar functions 6
HARDY 17, 19, 22, 29, 31, 34, 35, 37, 43, 44, 48
Hardy's inequality 37
HARTMAN 8
HEYWOOD 20, 31, 43, 55, 56, 59
HYLTÉN-CAVALLIUS 12

IGARI 56
Integrals §9
IZUMI, M. 22, 29, 55, 56
IZUMI, S. 22, 28, 29, 32, 55, 56

Jacobi polynomials 6

KOKILAŠVILI 6
KONYUŠKOV 45

L^p theorems 4, 5, §6, 56
$L(q,p)$ 35
Λ_* 46, 47
$\Lambda(\alpha, p, q)$ 46
Lacunary series 31, 32
LANDAU 12
Lebesgue constants 27
LEINDLER 6, 45
Lipschitz conditions §7, 59
LITTLEWOOD 17, 29, 31, 34, 35, 37
LORENTZ 35, 45
LUKE 10

MACAEV 36
MORAN 10

OFFORD 33
O'SHEA 19, 20, 23, 54

PAK 8
PALEY 34, 46, 48
Parseval's theorem 3, 6, 53, 54, 55, 57, 58
Partial summation 7
PEYERIMHOFF 52
PÓLYA 37
Positive coefficients 4, 5, §§4–6, 54
Positive functions 4, §3, 26, 29
Positive trigonometric sums 8, 9, 12

Quasiconvex 27
Quasimonotonic 6, 23, 54
Quasipositive 54

REES 46
Riemann summability 34
ROBERTSON 20, 37, 55, 59
ROGOSINSKI 22, 29, 34, 43, 48

SATÔ, M. (M. IZUMI) 22, 55
SHAH 23, 54
Sine transforms §9
Slowly increasing 52, 55
Smooth functions 47
Special trigonometric series 3, 10, 11, 12

ŠNEĬDER 6
STEČKIN 29
Summable series 28, 34
SUNOUCHI 19
SZÁSZ 34, 46
SZIDON 26, 29
SZ.-NAGY 13, 57

TELJAKOVSKIĬ 6, 27
TOMIĆ 8, 12, 43, 45, 52
TURÁN 12

Ultimately 1
Ultraspherical polynomials 6
Uniform convergence 47

WAINGER 4, 6, 19, 35, 54, 59
Walsh functions 6
Weierstrass function 47
WEISS 31
WINTNER 8

YANG 52
YONG 54, 56
YOUNG 19, 26

ZYGMUND 8, 10, 13, 14, 22, 26, 29, 30, 33, 35, 37, 43, 46, 48

Druck: Zechnersche Buchdruckerei, Speyer